# Mathematics is Beautiful

Heinz Klaus Strick

# Mathematics is Beautiful

Suggestions for People Between 9 and 99 Years to Look at and Explore

 Springer

Heinz Klaus Strick
Leverkusen, Germany

The translation was done with the help of artificial intelligence (machine translation by the service DeepL.com). A subsequent human revision was done primarily in terms of content.

ISBN 978-3-662-62688-7      ISBN 978-3-662-62689-4   (eBook)
https://doi.org/10.1007/978-3-662-62689-4

Responsible Editor: Iris Ruhmann
This Springer imprint is published by the registered company Springer-Verlag GmbH, DE part of Springer Nature.
The registered company address is: Heidelberger Platz 3, 14197 Berlin, Germany

# Preface

Not everyone thinks of mathematics as something to enjoy when talking about it. But mathematics has many exciting and aesthetically pleasing aspects to offer. In this book, I have tried to show some of these beautiful things in mathematics.

During my work as a mathematics teacher, I have always endeavored to loosen up my lessons to a certain extent. Unfortunately, even in the most exciting mathematics lessons tedious and dry phases cannot be avoided.

For such relaxation and enrichment, there are questions that could be classified as *mathematical games,* or even *brain teasers* whose solutions lead to amazing insights.

Thus, for example, after the treatment of the inscribed angle theorem in elementary geometry, regular star figures can be examined (Chap. 1) or regular polygons can be laid out using diamonds (Chap. 10). Searching the greatest common divisor of two numbers is more entertaining if one interprets this as the dissection of a rectangle (Chap. 3). Mental arithmetic is not to everyone's taste, but surprisingly, you can discover interesting structures in the world of numbers with just a few arithmetical tricks (Chap. 7). Solving quadratic equations and linear systems of equations is usually not very exciting – unless you use these methods to explore wonderful figures with touching circles *(Kissing circles,* Chap. 15) or to deal with the question of the tessellation of rectangles by squares of different sizes *(Squaring the square,* Chap. 14). In addition to the *Kissing circles* problems from Japanese temple geometry *(Sangaku)* are examined.

Several of the topics addressed in the book are aimed at younger students. Experience has shown that thread pictures *(Curve stitching,* Chap. 6) are extremely fascinating – even if the theoretical background can only be conveyed at the end of secondary school or even afterwards. Playing with pentominoes (Chap. 5) encourages a strategic and logical approach. And smart 10-year-olds can understand that weighing with a fixed, very limited set of balance weights (Chap. 9) conceals arithmetic in the ternary numeral system.

In the first years of school, children already learn to determine the areas of simple geometric figures; it is all the more astonishing, that a completely different way of measuring can be chosen: the area inside a polygon can be calculated when the vertices are points of a square-ruled paper: you only have to count the lattice points of the boundary and those lying inside the figure (Chap. 11). As an introduction to the subject this chapter also includes studies of rectangles and other simple figures on grid paper.

However, studying beautiful mathematics can also mean looking at colored patterns or designing one's own patterns. Patterns made of colored stones (Chap. 2) were already studied 2500 years ago. When coloring circular rings (Chap. 4) and equally large subareas of regular polygons (*Area divisions*, Chap. 8) you can develop your own imagination and perhaps even discover new patterns.

At the end of the book, there are two more extensive chapters on the derivation of power sum formulas (Chap. 16) and on the Pythagorean theorem (Chap. 17). They make clear how new ideas on a topic have been developed over the centuries.

Unfortunately, there was no room in this book for other topics. I am aware that a selection could have been different. (For example, if you miss the "Golden Ratio": at least some aspects can be found in Chaps. 3 and 13, but here will be a lot more in the second volume of "Mathematics is beautiful").

The chapters can be read independently of each other. At least when starting with the individual topics, the simplest possible approach was chosen; for this none or only a little background knowledge from school lessons is required.

It is an important concern of the book that – by reading this book – many young people find their way to mathematics and at the same time those readers, whose school days are some time ago, remember again and discover something new. The numerous references to further sources for information on the Internet as well as to further literature should help here. The "solutions" to the problems described in the individual sections *Suggestions for reflection and for investigations* are published on the author's website: https://www.mathematik-ist-schoen.de/mathematics-is-beautiful/.

This book was written for everyone who enjoys mathematics or wants to understand why the book bears this title. It is also aimed at teachers who want to give their students additional or new motivation to learn.

Even though each chapter contains – graphically emphasized – theorems, rules, and formulas, that is, the typical elements of a mathematics book, this is not a textbook of mathematics. Proofs of theorems are only based on examples – it was always more important to me to convey the underlying ideas than pointing out the formal conclusions.

The abundance of graphics in this book should encourage you to develop your own ideas about the objects presented:

*Viewing, thinking, trying out, varying, researching, wondering.*

The fact that most of the graphics were created using the LOGO programming language may be criticized, as the graphic resolution that can be achieved with this software is certainly not optimal. Besides the licensing issue, the decisive factor for my decision was my own positive teaching experiences with the concept of the programming language, which the inventor Seymour Papert (*Mindstorms*) himself considered suitable for primary school.

In recent years, I have had the pleasure of dealing with a new mathematician every month (https://www.spektrum.de/mathematik/monatskalender/index/). A lot of those "histories", which, with the help of John O'Connor, are now also available in English and can be downloaded from https://mathshistory.st-andrews.ac.uk/Strick/.

When you deal with the insights and ideas of scholars who have long since passed away, you often cannot help but be amazed. I hope that in this book I have also succeeded in bringing some of these wonderful insights, which have unfortunately often been forgotten, back into consciousness. I have made every effort to provide sufficient suggestions for further study of the topics by the literature references in each chapter and at the end of the book. Fortunately, the quality of the Wikipedia contributions (and the bibliographical references they each contain) has increased significantly in recent years. Sometimes they are even surpassed by the German or French version; therefore, these sources are also mentioned. It is no longer possible for me to state in detail which publications have given me which stimulus. Over the past decades I have worked through a large number of books, whose titles often begin with the words

*Recreations, Challenging Problems, Excursions, Adventures …*

Most of the time I looked at them from the point of view of whether they contained suggestions for "normal" lessons, for study groups, or as problems for competitions.

At the end of the work on this book, I would like to thank all those who have supported me in the preparation and implementation of the book project:

- To my wife, who patiently put up with the fact that I kept immersing myself in the beautiful world of mathematics,
- To Wilfried Herget (University of Halle), who made numerous suggestions to make the wording of my texts more understandable and revealed gaps in arguments,
- To Manfred Stern †(University of Halle), Peter Gallin (University of Zurich), and Hans Walser (University of Basel) who have given numerous suggestions for this book,
- To John O'Connor (University of St Andrews) who liberally helped so that this book could be published in an understandable translation,
- And not least to Andreas Rüdinger, Iris Ruhmann, Carola Lerch, Snehal Surwade and Jasmeen Kaur from Springer Verlag, who made this book possible.

Leverkusen Germany                                                    Heinz Klaus Strick

# Contents

# Regular Polygons and Stars

**1**

*Three things remain with us from paradise:*
*Stars, flowers and children.*

*(Dante Alighieri, 1265–1321, Italian poet and philosopher)*

## 1.1 Properties of Regular Stars

Regular stars are created by connecting vertices of regular polygons according to a certain rule.

Such a rule could be worded as follows:

Connect one vertex of the polygon with the $k$-next vertex (clockwise).

**Example: 5-Pointed Star (Pentagram)**
For $n = 5$ and $k = 2$, this means: connect each vertex of a regular 5-sided figure (pentagon) to the second-next vertex (clockwise). Thus a regular 5-pointed star is created.

© Springer-Verlag GmbH Germany, part of Springer Nature 2021
H. K. Strick, *Mathematics is Beautiful*, https://doi.org/10.1007/978-3-662-62689-4_1

No further 5-pointed stars exist, because for $n = 5$ and $k = 3$ you get the same star. Instead of connecting each vertex to the third-next vertex clockwise, you can connect the vertex to the second-next vertex counterclockwise.

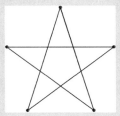

### Example: 6-Pointed Star (Hexagram)

Also for $n = 6$ only one type exists. It consists of two 3-sided figures (equilateral triangles), because $2 \cdot 3 = 6$.

If you number the vertices of the $n$-sided figure clockwise with $P_0, P_1, P_2, P_3, P_4, P_5$, then you get two closed polygonal lines: $P_0 - P_2 - P_4 - P_0$ and $P_1 - P_3 - P_5 - P_1$, with either even or odd indices.

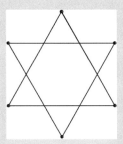

### Example: 7-Pointed Stars (Heptagrams)

For $n = 7$ there are two different stars, namely for $k = 2$ and for $k = 3$. If you look closely, you can see that the 7-pointed star for $k = 2$ is also created inside the star for $k = 3$ (also a regular 7-sided figure).

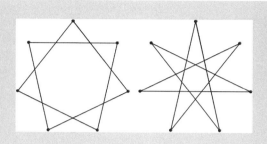

**Example: 8-Pointed Stars (Octagrams)**

Also for $n = 8$ there are two different stars, that is for $k = 2$ and for $k = 3$.

The 8-pointed star for $k = 2$ also appears inside the star for $k = 3$. It consists of two regular 4-sided figures (squares), because $2 \cdot 4 = 8$.

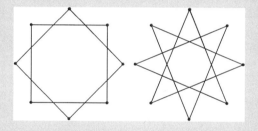

**Example: 9-Pointed Stars (Enneagrams)**

For $n = 9$ there are even three different stars.

- $n = 9, k = 2$: The star can be drawn as a closed polygonal line:

$$P_0 - P_2 - P_4 - P_6 - P_8 - P_1 - P_3 - P_5 - P_7 - P_0$$

- $n = 9, k = 3$: The star consists of three regular 3-sided figures (equilateral triangles), because $3 \cdot 3 = 9$.
- $n = 9, k = 4$: The star can be drawn as a closed polygonal line:

$$P_0 - P_4 - P_8 - P_3 - P_7 - P_2 - P_6 - P_1 - P_5 - P_0$$

Inside, the stars for both, $k = 2$ and $k = 3$, appear.

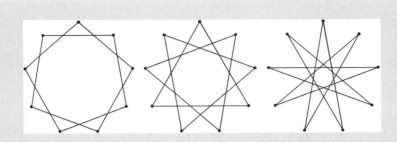

**Example: 10-Pointed Stars (Decagrams)**
There are also three different stars for $n = 10$.

- $n = 10$, $k = 2$: This star consists of two regular 5-sided figures, because $2 \cdot 5 = 10$.
- $n = 10$, $k = 3$: The star can be drawn as a closed polygonal line.
- $n = 10$, $k = 4$: This star consists of two stars of type $n = 5$, $k = 2$. These include the two closed polygonal lines $P_0 - P_4 - P_8 - P_2 - P_6 - P_0$ and $P_1 - P_5 - P_9 - P_3 - P_7 - P_1$.

**Example: 11-Pointed Stars (Hendecagrams)**
For $n = 11$ there are four different stars, namely for $k = 2$, $k = 3$, $k = 4$, and $k = 5$.
  All of these stars can be drawn as closed polygonal lines.
  On the inside the stars with smaller $k$ appear respectively.

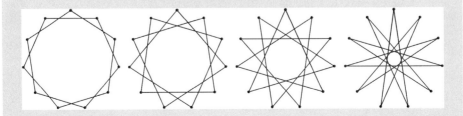

**Example: 12-Pointed Stars (Dodecagrams)**
For $n = 12$ there are four different stars:

- $k = 2$: 2 regular 6-sided figures, because $2 \cdot 6 = 12$.
- $k = 3$: 3 regular 4-sided figures (squares), because $3 \cdot 4 = 12$.
- $k = 4$: 4 regular 3-sided figures (equilateral triangles), because $4 \cdot 3 = 12$.

Only the star for $k = 5$ can be drawn as a closed polygonal line.
On the inside the stars with smaller $k$ appear respectively.

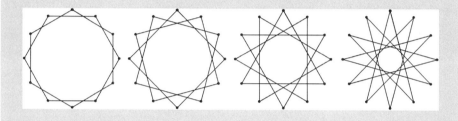

The following properties can be identified from the examples:

- $n$-pointed stars exist for every $n$, which is greater than 4.
- For $k$ you can use any number. You can get different star figures, if you use the following values in the drawing rule: $k$ is at least 2, for even-numbered $n$ use at most $\frac{n}{2} - 1$, for odd-numbered $n$ use at most $\frac{n-1}{2}$.
  - In detail, the following applies for odd-numbered $n$: for $n = 5$ there is one star for $k = 2$; for n $= 7$ there are two stars, namely for $k = 2$ and for $k = 3$; for $n = 9$ there are three stars, namely for $k = 2$ for $k = 3$ and for $k = 4$; and so on.
  - In detail, the following applies for even-numbered $n$: for $n = 6$ there is one star for $k = 2$; for $n = 8$ there are two stars, namely for $k = 2$ and for $k = 3$; for $n = 10$ there are three stars, namely for $k = 2$, for $k = 3$ and for $k = 4$; and so on.
- If any vertex is determined as the beginning of a closed polygonal line with the number 0, then the line passes through the vertices with the numbers $0 - k - 2k - 3k - \cdots$, and similar as to a clock, the numbers are each reduced by $n$, when the multiple of $k$ reaches or exceeds the number $n$.
- In every $n$-pointed star, there are further $n$-pointed stars inside for every possible $k > 2$.
- Some star figures can be drawn without lifting the pen; others consist of two or more polygons or star figures. In detail:
  - If $k$ is a divisor of $n$, then the star consists of $k$ polygons with $e$ vertices, where $e = \frac{n}{k}$.

- If $k$ and $n$ have the common divisor $g$, then the $n$-pointed star is composed of $g$ stars with $\frac{n}{g}$ vertices.
- If $k$ and $n$ are coprime, that is, if they only have the number 1 as a common divisor, the star can be drawn as a (single) closed polygonal line. Conversely, if a star can be drawn as a (single) closed polygonal line, then $k$ and $n$ are coprime.

---

**Rule**

**Stars that can be Drawn as a Closed Polygonal Line**

Regular $n$-pointed stars exist for all natural numbers $n$, $k$ with $n > 4$ and $2 \leq k \leq \frac{n}{2} - 1$, if $n$ is an even number, or $2 \leq k \leq \frac{n-1}{2}$, if $n$ is an odd number.

Then, and only then, the stars can be drawn as a closed polygonal line, if $n$ and $k$ are coprime. ◄

Since in regular $n$-pointed stars both the number of vertices $n$ and the parameter $k$ play an important role, they are often notated with the symbolic notation $\{n/k\}$, the so-called **Schläfli symbol** (named after the Swiss mathematician Ludwig Schläfli [1814–1895], who was particularly interested in regular polygons, polyhedrons and their generalization in higher dimensions).

**Suggestions for Reflection and for Investigations**

**A 1.1:** Answer the following questions for $n = 13$, $n = 15$, and for $n = 18$ (that is, for an odd or even number of vertices): for which $k$ (minimum and maximum value) do you get an $n$-pointed star? How many different star figures are possible? Which of the possible star figures can be drawn as a closed polygonal line, which consist of several stars, which of several polygons? Which numbers of vertices appear in the possible closed polygonal lines (start of lines at the vertex with number 0)?

**A 1.2:** In the following figures, areas of equal size are colored in the same way. How does the number of colors depend on the type of star, i.e. on the values for $n$ and $k$?

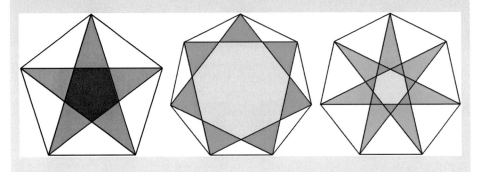

## 1.2   Drawing Stars

To draw a regular star with $n$ vertices, you need to know how to draw a regular $n$-sided polygon.

Especially simple is the construction of a regular 4-sided figure (square) and a regular 6-sided figure (hexagon) as well as the regular polygons, each obtained by doubling the number of vertices from given regular $n$-sided figures:

- A regular 4-sided figure is obtained by drawing a circle of any radius $r$, selecting any point on the circle and drawing a straight line through the center of the circle until the circular line is intersected again. Then draw a perpendicular to this line through the center of the circle to get two more points of the 4-sided figure. These four points determine a square.
- A regular 6-sided figure is created by drawing a circle with an arbitrarily chosen radius $r$, then selecting any point on the circular line and from this point successively drawing lines of the length $r$ on the circle. This construction is possible because the regular 6-sided figure consists of six equilateral triangles, i.e., the sides of the 6-sided figure are as long as the line segments which connect the vertices with the center of the circle (= radius of the circle).

If you draw a straight line from the center of the circle through each of the centers of the sides of the regular $n$-sided polygon, then the intersection points of these straight lines with the circular line are the additional vertices for the regular $2n$-sided polygon. In this way you will get out of the square a the regular 8-sided polygon, from the regular 6-sided polygon you will get the regular 12-sided polygon, and so on (see the following figures).

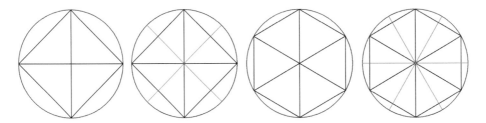

In general, that is, for any *n,* there are two possibilities:

- You start with a circle with radius *r,* which is drawn around a center point, and then draw the radius *n*-times from the center, changing the direction 360°/*n* each time.
  Figure 1.1 shows (for *n* = 7) not only the vertices but also the sides of the regular *n*-sided polygon and the altitudes of the resulting isosceles triangles. The *n*-pointed star is created when a starting point is connected with the *k*-next point according to the rules, and this procedure is then repeated *n* times.
- Alternatively, you can also start with one side of the *n*-sided polygon, that is, draw a line of length *s,* then change the direction in which you moved while drawing by the *n*th part of 360°, so that after repeating the process *n* times, you have made a total rotation of 360° and have arrived back at the starting point of the "walking tour."

There is a simple relationship between the circle radius *r* and the side length *s* of the regular *n*-sided polygon: two adjacent radii and one side of the *n*-sided polygon form an isosceles triangle, which is divided by the altitude *h* into two right-angled triangles.

Therefore, the following applies to the half angle at the center:

$$\sin\left(\frac{180°}{n}\right) = \frac{s}{2r} \text{ and } \tan\left(\frac{180°}{n}\right) = \frac{s}{2h} \text{ and } \cos\left(\frac{180°}{n}\right) = \frac{h}{r}$$

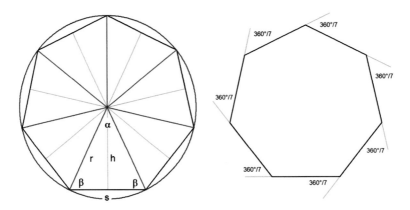

**Fig. 1.1**   Two of the ways to draw a regular 7-sided polygon

## 1.3   Diagonals in a Regular *n*-Sided Figure

In exploring the question which *n*-pointed stars are possible at all, it makes sense to draw a regular *n*-sided figure with all diagonals first and then, according to the instructions, mark the desired closed polygonal line for which the diagonals are used.

From each vertex of an *n*-sided figure you can draw line segments to the other vertices: 2 sides (to the two adjacent vertices) and $n - 3$ diagonals (to the remaining vertices).

The total number of diagonals in an *n*-sided polygon does not result directly from the product $n \cdot (n - 3)$ because with this method of counting each of the connecting lines is counted twice. Rather the following applies:

---

**Rule**

**Number of Diagonals of an *n*-Sided Polygon**

The number of diagonals in an *n*-sided polygon is equal to $\frac{1}{2} \cdot n \cdot (n - 3)$. ◀

---

**Examples for the Calculation of the Number of Diagonals**

A regular 5-sided figure has $\frac{1}{2} \cdot 5 \cdot 2 = 5$ diagonals that form the regular 5-pointed star.

A regular 6-sided figure has $\frac{1}{2} \cdot 6 \cdot 3 = 9$ diagonals, but 3 of them only lead to the opposite point, so they are not suitable to draw a star. The remaining 6 diagonals form the 3 sides of the two equilateral triangles.

A regular 7-sided figure has $\frac{1}{2} \cdot 7 \cdot 4 = 14$ diagonals, of which 7 diagonals each form a polygonal line for the 7-pointed star with $k = 2$ or $k = 3$.

A regular 8-sided figure has $\frac{1}{2} \cdot 8 \cdot 5 = 20$ diagonals, of which 4 only lead to the opposite point, so they are not suitable to draw a star. In addition, two times four diagonals each form the two squares of which star {8/2} consists, so that 8 diagonals remain, which form the regular 8-pointed star {8/3}.

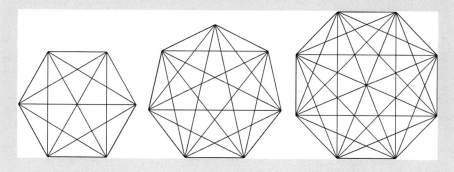

**Suggestions for Reflection and for Investigations**
**A 1.3:** Determine the number of diagonals for $n = 9$ to $n = 12$ in the regular $n$-sided polygon. Which of these diagonals are needed for drawing $n$-pointed stars? Generalize these statements about diagonals and stars for an even and odd number of vertices.

In the regular 5-sided figure (pentagon), all diagonals have the same length. If you connect the end points of a diagonal to the center of the circle, an isosceles triangle with base $d$ and two legs of the length $r$ is formed. Since the diagonals connect one vertex of the regular 5-sided figure with the second next vertex, the size of the angle $\delta$ at the center of the circle is equal to $2 \cdot \frac{360°}{5}$ that is, the size of half the angle is equal to $2 \cdot \frac{180°}{5} = 72°$.

Therefore applies to the diagonals in the regular 5-sided figure:

$$\sin\left(\frac{2 \cdot 180°}{5}\right) = \frac{\frac{d}{2}}{r}, \text{ that is } d = 2r \cdot \sin\left(\frac{2 \cdot 180°}{5}\right).$$

In general, for the diagonals in any regular $n$-sided polygon, which connect one vertex with the second next vertex, the length of the diagonal $d_2$ is given as:

$$d_2 = 2r \cdot \sin\left(\frac{2 \cdot 180°}{n}\right)$$

In the case of diagonals connecting one vertex with the third next vertex, the angle $\delta$ at the center of an isosceles triangle changes accordingly to $3 \cdot \frac{360°}{n}$, that is, half the angle to $3 \cdot \frac{180°}{n}$. Therefore, the following applies:

$$d_3 = 2r \cdot \sin\left(\frac{3 \cdot 180°}{n}\right)$$

---

**Formula**
**Length of the Diagonals of a Regular $n$-Sided Polygon**
In general, for the length $d_k$ of a diagonal, that connects a vertex with the $k$-next vertex of a regular $n$-sided polygon and that lies opposite to the angle $\delta = k \cdot \frac{360°}{n}$, the following applies:

$$d_k = 2r \cdot \sin\left(\frac{k \cdot 180°}{n}\right) \tag{1.1}$$

By means of formula (1.1), the total length of the closed polygonal line which forms the regular $n$-pointed star can then be calculated, see also Table 1.1 below. ◀

**Table 1.1** Angular sizes and line lengths for regular *n*-pointed stars

| Star type {*n/k*} | Number of polygonal lines | Center angle $\delta_k$ (opposite to the diagonal $d_k$) $\delta_k = k \cdot \frac{360°}{n}$ (°) | Angle $\varepsilon$ at the "tip" (°) | Total length of all lines of the star $n \cdot 2r \cdot \sin\left(\frac{k \cdot 180°}{n}\right)$ |
|---|---|---|---|---|
| {5/2} | 1 | 144 | 36 | 9.51 · r |
| {6/2} | 2 | 120 | 60 | 10.39 · r |
| {7/2} | 1 | 102.86 | 77.14 | 10.95 · r |
| {7/3} | 1 | 154.29 | 25.71 | 13.65 · r |
| {8/2} | 2 | 90 | 90 | 11.31 · r |
| {8/3} | 1 | 135 | 45 | 14.78 · r |
| {9/2} | 1 | 80 | 100 | 11.57 · r |
| {9/3} | 3 | 120 | 60 | 15.59 · r |
| {9/4} | 1 | 160 | 20 | 17.73 · r |
| {10/2} | 2 | 72 | 108 | 11.76 · r |
| {10/3} | 1 | 108 | 72 | 16.18 · r |
| {10/4} | 2 | 144 | 36 | 19.02 · r |
| {11/2} | 1 | 65.45 | 114.55 | 11.89 · r |
| {11/3} | 1 | 98.18 | 81.82 | 16.63 · r |
| {11/4} | 1 | 130.91 | 49.91 | 20.01 · r |
| {11/5} | 1 | 163.64 | 16.36 | 21.78 · r |
| {12/2} | 2 | 60 | 120 | 12 · r |
| {12/3} | 3 | 90 | 90 | 16.97 · r |
| {12/4} | 4 | 120 | 60 | 20.78 · r |
| {12/5} | 1 | 150 | 30 | 23.18 · r |

## 1.4 Vertex Angle in a Regular *n*-Pointed Star

At the vertices of the regular *n*-pointed stars, there are angles that depend on the values for *n* and *k*. These are easy to determine by applying the so-called **inscribed angle theorem**. The theorem deals with the central angle above a chord and the associated inscribed angle (peripheral angle) above it. The theorem states that all peripheral angles above a chord are equal. The central angle is twice as large as the periphal angles.

Figure 1.2 shows the symmetric case of the theorem; for a general proof of the theorem look at the references.

If two adjacent vertices of a regular *n*-sided figure are connected to each other, then the central angle belonging to the side of the *n*-sided figure is equal to $\frac{360°}{n}$; the corresponding peripheral angles are equal to $\frac{180°}{n}$.

**Fig. 1.2** Relationship between
the center angle and the
peripheral angle in a symmetric
triangle

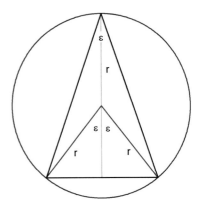

If you connect a vertex of a regular $n$-sided figure with the second next vertex, then the central angle belonging to this diagonal $d_2$ is twice as large as $\frac{360°}{n}$ thus equal to $\frac{720°}{n}$ and the corresponding peripheral angles are equal to $\frac{360°}{n}$.

In general:

---

**Rule**

**Central Angles and Peripheral Angles Over a Chord in Regular $n$-Sided Polygons**

If you connect a vertex of a regular $n$-sided polygon with the $k$-next vertex, then the angle at the center of this diagonal $d_k$ is $k$-times as big as $\frac{360°}{n}$; the corresponding peripheral angles are equal to $k \cdot \frac{180°}{n}$. ◀

---

**Examples of the Angles in the Vertices of Regular n-Pointed Stars**

- With the regular 5-pointed star the vertex is "above" one side of the 5-sided figure. Therefore, the angle $\varepsilon$ at the vertex is half the angle at the center of the regular 5-sided figure. Since the angle at the center has an angular size of $\frac{360°}{5} = 72°$, the angle at the vertex of the regular 5-pointed star is $\varepsilon = \frac{180°}{5} = 36°$ - see the first of the following figures.

- In the regular 6-pointed star, the vertex is also "above" a diagonal of the 6-sided figure, which connects one vertex with the second-next. Therefore the angle $\varepsilon$ is half as large as the corresponding central angle, that is, half as large as $2 \cdot \frac{360°}{6}$, that is $\varepsilon = 60°$, see the second of the following figures.

- With the regular 7-pointed star $\{7/2\}$ the vertex is also "above" a diagonal of the 7-sided figure, which connects one vertex with the third next vertex. Therefore, the angle $\varepsilon$ is half as large as the corresponding central angle, namely half the size of $3 \cdot \frac{360°}{7}$, that is $\varepsilon \approx 77.14°$.

On the other hand, with the star {7/3} the point is "above" a diagonal of the 7-sided figure, which connects one vertex with the next vertex. Therefore, the point angle $\varepsilon$ is half as large as the corresponding central angle, namely half as large as $1 \cdot \frac{360°}{7}$, that is $\varepsilon \approx 25.71°$, see the third and fourth of the following figures.

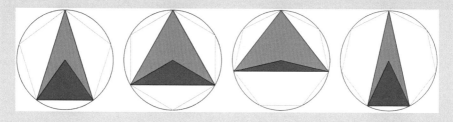

**Suggestions for Reflection and for Investigations**

**A 1.4:** Using the 8-, 9-, 10-, or 12-pointed stars shown in the figure, consider which are the angular sizes in the vertices of the *n*-pointed stars.

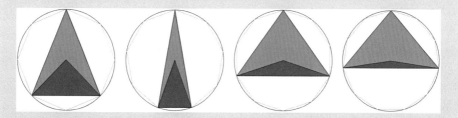

**A 1.5:** One of the regular 9-pointed stars has a central angle greater than 180°. Use the following two figures to explain how the angle in the vertex is calculated here.

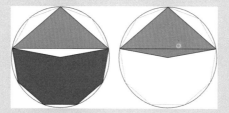

**A 1.6:** The following regular stars also have a central angle that is greater than 180°. In each case, explain how the angles in the vertices are calculated.

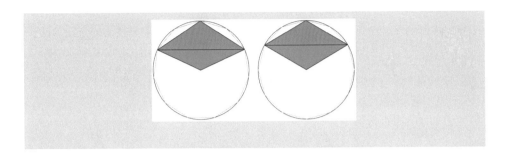

On the basis of the examples, it can be assumed that there is a simple relationship between the angle $\varepsilon$ in the vertex and the angle at the center $\delta_k$ above the diagonals, namely $\varepsilon = 180° - \delta_k$, see the following table.

| Star type {n/k} | Center angle $\delta_k$ (opposite to the diagonal $d_k$) | Angle $\varepsilon$ (at the "tip") |
|---|---|---|
| {5/2} | 144° | 36° |
| {6/2} | 120° | 60° |
| {7/2} | 102.86° | 77.14° |
| {7/3} | 154.29° | 25.71° |

Figure 1.3 shows that this is true: the vertex is determined by two diagonals, of which each has the central angle $\delta_k$. According to Sect. 1.3 this angle can be calculated as $\delta_k = k \cdot \frac{360°}{n}$. For the base angles $\gamma$ of the associated isosceles triangles, the following applies, due to the angle sum in the triangle, $2\gamma + \delta_k = 180°$.

But since the vertex angle $\varepsilon$ consists of twice the angle $\gamma$, the proposition applies $\varepsilon + \delta_k = 180°$.

**Fig. 1.3** To determine the angle $\varepsilon = 2\gamma$ at the vertex of a regular $n$-sided polygon

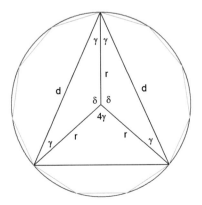

---

**Rule**

**Size of the Vertex Angles in Regular *n*-Pointed Stars**

For the vertex angle $\varepsilon$ of a regular *n*-pointed star of the type $\{n/k\}$ the following applies:

$$\varepsilon = 180^\circ - \frac{k \cdot 360^\circ}{n}$$

Inside a star of type $\{n/k\}$ further *n*-pointed stars $\{n/m\}$ appear with $1 < m < k$. At the very center of a regular star there is also a regular *n*-sided figure, for whose interior angles $\alpha$ applies: $\alpha = 180^\circ - \frac{360^\circ}{n}$.

So you can apply the formula for calculating $\varepsilon$ also to the case $k = 1$ and mark regular *n*-sided figures with the Schläfli symbol $\{n/1\}$.

The results so far are shown in Table 1.1. ◀

---

## 1.5 Compounded *n*-Pointed Stars

In principle, you can also create regular *n*-pointed stars by first creating a regular *n*-sided polygon, and then drawing isosceles triangles above the sides of the polygon. In the following figures, equilateral and *golden* triangles, respectively have been placed on the sides of a regular 5, 6, and 7-sided figure. (Isosceles triangles with a base angle of 72° are called golden triangles).

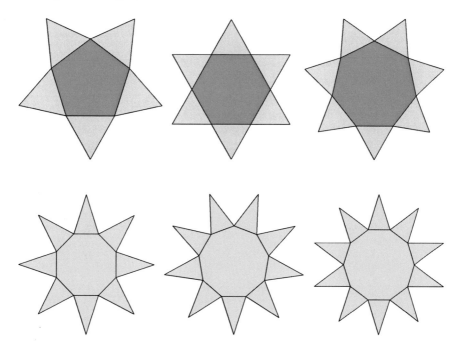

**Suggestions for Reflection and for Investigations**

**A 1.7:** Prove the proposition: all regular $n$-pointed stars of the type $\{n/2\}$ can be interpreted as compounded $n$-pointed stars.

## 1.6     Regular $n$-Sided Figures in the Complex Plane

Section 1.2 explained how to draw regular $n$-sided polygons. No coordinate system is required for these drawings.

In complex analysis, one often uses representations based on the so-called **complex plane** (also called "Argand diagram" named after the French amateur mathematician Jean-Robert Argand, 1768–1822). This is a two-dimensional coordinate system in which the real part of a complex number is plotted in horizontal direction and the imaginary part in vertical direction.

Complex numbers $z = x + i \cdot y$ are defined in the coordinate system of the complex plane as points with the coordinates $(x, y)$ (see Fig. 1.4).

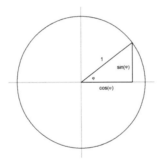

The points $(x, y)$ of the unit circle, that is, a circle with the radius 1, satisfy the equation $x^2 + y^2 = 1$ – according to the Pythagorean theorem. If you name  the angle between the ray, leading from the center to a point on the unit circle, and the $x$-axis with $\varphi$, then each point can also be described by the coordinates $(\cos(\varphi), \sin(\varphi))$.

**Fig. 1.4**  Stamps of the postal service of the Federal Republic of Germany ("Deutsche Bundespost") on C. F. Gauss and the complex plane (in Germany named as "Gauss'sche Zahlenebene")

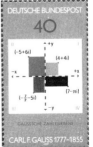

An equation of the form $z^n = 1$ is called as **cyclotomic polynomial equation**. According to the the **fundamental theorem of algebra**, such an equation has exactly $n$ solutions in the set of complex numbers. In the complex plane, the solutions of the cyclotomic polynomial equation form the vertices of a regular $n$-sided figure (hence the name for the equation).

The French mathematician Abraham de Moivre (1667–1754), who lived in exile in England, discovered that for every complex number $z = \cos(\varphi) + i \cdot \sin(\varphi)$ and for each natural number $n$ the following equation applies:

---

**Formula**

**Theorem of Moivre**

$$[\cos(\varphi) + i \cdot \sin(\varphi)]^n = \cos(n \cdot \varphi) + i \cdot \sin(n \cdot \varphi) \tag{1.2}$$

Therefore the following applies for every angle $\varphi = k \cdot \frac{360°}{n}$ with $k = 0, 1, 2, \ldots, n - 1$:

$$\left[ \cos\left( k \cdot \frac{360°}{n} \right) + i \cdot \sin\left( k \cdot \frac{360°}{n} \right) \right]^n = \cos(k \cdot 360°) + i \cdot \sin(k \cdot 360°) = 1 + i \cdot 0 = 1$$

That means that the $n$ complex numbers $z_k = \cos\left( k \cdot \frac{360°}{n} \right) + i \cdot \sin\left( k \cdot \frac{360°}{n} \right)$ satisfy the equation $z^n = 1$. ◀

---

**Formula**

**Solutions of the Cyclotomic Polynomial Equation**

The $n$ solutions of the cyclotomic polynomial equation $z^n = 1$ have the form

$$z_k = \cos\left( k \cdot \frac{360°}{n} \right) + i \cdot \sin\left( k \cdot \frac{360°}{n} \right),$$

where $k = 0, 1, 2, \ldots, n - 1$.

The $n$ solutions can be found by drawing $n$ rays from the origin of the coordinate system with an angle $\varphi$ with $\varphi = k \cdot \frac{360°}{n}$ and determining their points of intersection with the unit circle.

In special cases, the solutions of the cyclotomic polynomial equation can also be determined using elementary algebraic methods, that is, without using trigonometric functions. This is illustrated by the examples for $n = 3$, 4, and 5. ◀

---

**Example 1: Solution of the Equation $x^3 = 1$**

**Using Trigonometric Functions:**

The cubic equation $z^3 = 1$ has the three solutions:

$$z_0 = \cos(0°) + i \cdot \sin(0°) = 1$$
$$z_1 = \cos(120°) + i \cdot \sin(120°) = -\tfrac{1}{2} + i \cdot \tfrac{\sqrt{3}}{2}$$
$$z_2 = \cos(240°) + i \cdot \sin(240°) = -\tfrac{1}{2} - i \cdot \tfrac{\sqrt{3}}{2}$$

**Using Algebraic Methods:**

The cubic equation $x^3 = 1$ has only one real-valued solution, namely $x_1 = 1$. This solution is represented in the complex plane by the point $(1, 0)$.

Since $x_1 = 1$ is a solution, the division of terms $\frac{x^3-1}{x-1}$ can be performed without remainder. This leads to the quadratic equation

$$x^2 + x + 1 = 0 \Leftrightarrow \left(x + \tfrac{1}{2}\right)^2 = -\tfrac{3}{4},$$

which has two complex solutions, namely

$$x_2 = -\tfrac{1}{2} + i \cdot \tfrac{\sqrt{3}}{2} \text{ and } x_3 = -\tfrac{1}{2} - i \cdot \tfrac{\sqrt{3}}{2}.$$

In the complex plane, these two solutions are drawn as points with the coordinates $\left(-\tfrac{1}{2}, \tfrac{\sqrt{3}}{2}\right)$ and $\left(-\tfrac{1}{2}, -\tfrac{\sqrt{3}}{2}\right)$.

**Example 2: Solution of the Equation x⁴=1**

**Using Trigonometric Functions:**

The 4th degree equation $z^4 = 1$ has the four solutions:

$$z_0 = \cos(0°) + i \cdot \sin(0°) = 1$$
$$z_1 = \cos(90°) + i \cdot \sin(90°) = i$$
$$z_2 = \cos(180°) + i \cdot \sin(180°) = -1$$
$$z_3 = \cos(270°) + i \cdot \sin(270°) = -i$$

**Using Algebraic Methods:**

The 4th degree equation $x^4 = 1$ has two real-valued solutions, namely $x_1 = 1$ and $x_2 = -1$. These solutions are represented in the complex plane by the points $(1, 0)$ and $(-1, 0)$.

Since $x_1 = 1$ and $x_2 = -1$ are solutions, the division of terms $\frac{x^4-1}{x^2-1}$ can be performed without remainder.

This leads to the quadratic equation $x^2 + 1 = 0$, which has two complex solutions: $x_3 = i$ and $x_4 = -i$.

In the complex plane, these two solutions are drawn as points with the coordinates $(0, 1)$ and $(0, -1)$.

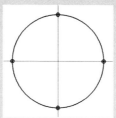

### Example 3: Solution of the Equation $x^5 = 1$

**Using Trigonometric Functions:**

The 5th degree equation $z^5 = 1$ has five solutions:

$$z_0 = \cos(0°) + i \cdot \sin(0°) = 1$$
$$z_1 = \cos(72°) + i \cdot \sin(72°)$$
$$z_2 = \cos(144°) + i \cdot \sin(144°)$$
$$z_3 = \cos(216°) + i \cdot \sin(216°)$$
$$z_4 = \cos(288°) + i \cdot \sin(288°)$$

**Using Algebraic Methods:**

The equation of the 5th degree $x^5 = 1$ has only one real-valued solution, namely $x_1 = 1$. This solution is represented in the complex plane by the point $(1, 0)$.

Since $x_1 = 1$ is a solution, the division of terms $\frac{x^5 - 1}{x - 1}$ can be performed without remainder.

This leads to the equation of the 4th degree $x^4 + x^3 + x^2 + x + 1 = 0$.

It can be shown that the polynomial can be factorized, namely

$$x^4 + x^3 + x^2 + x + 1 = \left(x^2 - \tfrac{1}{2} \cdot \left(\sqrt{5} - 1\right) \cdot x + 1\right) \cdot \left(x^2 + \tfrac{1}{2} \cdot \left(\sqrt{5} - 1\right) \cdot x + 1\right).$$

So we get four complex solutions of the two associated quadratic equations.

**Suggestions for Reflection and for Investigations**

**A 1.8:** Explain the required steps

$$x^6 - 1 = (x^2 - 1) \cdot (x^4 + x^2 + 1) = (x^2 - 1) \cdot (x^2 + x + 1) \cdot (x^2 - x + 1)$$

and use it to determine the coordinates of the corresponding points in the complex plane, which define a regular 6-sided figure.

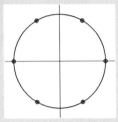

**A 1.9:** Explain the required steps

$$x^8 - 1 = (x^2 - 1) \cdot (x^6 + x^4 + x^2 + 1)$$
$$= (x^2 - 1) \cdot (x^2 + 1) \cdot (x^2 + \sqrt{2}x + 1) \cdot (x^2 - \sqrt{2}x + 1)$$

and use it to determine the coordinates of the corresponding points in the complex plane, which define a regular 8-sided figure.

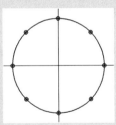

Relatively few cyclotomic polynomial equations can be solved using algebraic methods: these are exactly those equations in which a construction of the associated regular $n$-sided figure is possible with the aid of a compass and a ruler.

For example, the cyclotomic polynomial equation $x^7 = 1$ and $x^9 = 1$ can be solved graphically, but not algebraically, whereas this is possible for the equations $x^{10} = 1$ and $x^{12} = 1$ (see following figures).

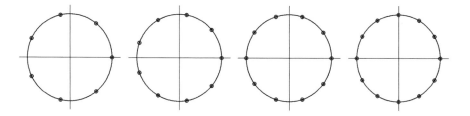

## 1.7    Setting up Game Schedules Using Regular *n*-Sided Figures

The structures of regular *n*-sided figures with their sides and diagonals can be used to set up game schedules for tournaments and  similar competitions.

**Example: Setting up Game Schedules for a Tournament with Eight Teams**
For a tournament with eight participating teams, a game schedule must be set up. Since each team has to play against every other team, seven rounds of four games each are to be played, making a total of 28 games.

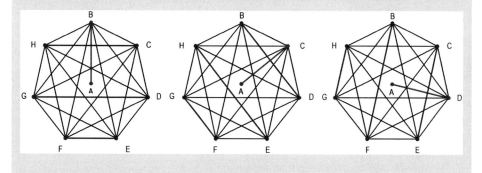

**Solution of the Problem**
Consider a regular 7-sided figure. This has 7 sides and $\frac{1}{2} \cdot 7 \cdot 4 = 14$ diagonals.

In addition, 7 radii play a role, which can be drawn from the vertices to the center, a total of 28 line segments.

The center A and the seven vertices B, C, D, E, F, G, H stand for the eight teams. The four games of a matchday are each determined by the radius marked in red and the three diagonals perpendicular to it and the side parallel to it (all marked in blue).

The fixtures of the seven rounds are obtained by looking at the seven different radii.

This results in the following games for the first rounds:

1. Round: A − B, C − H, D − G, E − F
2. Round: A − C, B − D, E − H, F − G
3. Round: A − D, C − E, B − F, G − H

**Suggestions for Reflection and for Investigations**
**A 1.10:** Complete the game schedule for the tournament with eight teams.
**A 1.11:** Create a game schedule for a tournament with seven teams.
**A 1.12:** Billiards is a game where two people or two teams play against each other. The so-called cue is used to push the balls on a special table covered with baize. The player only hits the white ball with the cue, which in turn can then hit other balls.

Five friends (A, B, C, D, E) regularly play billiards together, always in teams of two, one player sitting out per round.

- How many games are played at a time until every possible two-man team has played against every possible other two-man team?
- Set up a game schedule in which players A, B, C, D and E hang up one after the other.

*Tip:* Use a regular pentagon with diagonals to create a game plan.

## 1.8    References to Further Literature

On **Wikipedia** you can find further information and literature on the keywords in English (German, French):

- Star polygon (Stern, Polygon régulier étoile),
- Schläfli symbol (Schläfli-Symbol, Symbole de Schläfli)
- List of regular polytopes and compounds (*in English only*)
- Cyclotomic polynomial (Kreisteilung/Kreisteilungspolynom, Polynôme cyclotomique)

More informations can be found at **Wolfram Mathworld** under the keywords:

- Star polygon, polygram and special cases such as pentagram, hexagram, cyclotomic polynomial, root of unity

# Patterns of Colored Stones

# 2

*Number rules the universe. Number is the within of all things.*
*Number is the ruler of forms and ideas, and the cause of gods and*
*demons.*

*(Pythagoras of Samos, 570–500 B.C.)*

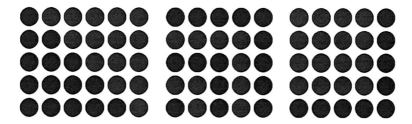

In the fifth century B.C. the Pythagoreans were already concerned with the question of how to illustrate mathematical laws. For this they used differently colored or light and dark stones.

## 2.1    Sum of the First *n* Natural Numbers

The simplest pattern is to lay out stones in a triangular shape, with the number of stones decreasing by 1 from top to bottom.

© Springer-Verlag GmbH Germany, part of Springer Nature 2021
H. K. Strick, *Mathematics is Beautiful*, https://doi.org/10.1007/978-3-662-62689-4_2

**Example: Determine the Sum 1 + 2 + 3 + ... + 10**

For each summand you place a corresponding number of (blue) stones in the form of an isosceles right-angled triangle (from top to bottom and left-aligned) one below the other.

Then double the figure and place the same number of red stones from bottom to top.

You get a rectangular pattern with a total of $10 \cdot 11 = 110$ stones.

The sum you are looking for is half as much, so the following applies:

$$1 + 2 + 3 + \ldots + 10 = 55$$

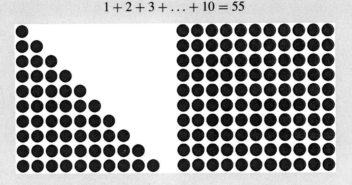

This can be generalized for the sum of any number $n$ of natural numbers:

To calculate the sum of the first $n$ natural numbers, you have to double the corresponding triangular figure. For this you need twice as many stones, so $2 \cdot (1 + 2 + 3 + \ldots + n)$, so that a rectangle of stones results, which has the height $n$ and the width $n + 1$.

Therefore, the following applies:

**Formula**

**Sum of the First $n$ Natural Numbers**

The sequence of the first $n$ natural numbers can be represented as an isosceles right-angled triangle.

The sum $(1 + 2 + 3 + \ldots + n)$ of the first $n$ natural numbers can be determined by multiplying the largest summand ($n$) with its successor ($n + 1$) and halving the product of the two numbers.

$$1 + 2 + \ldots + n = \tfrac{1}{2} \cdot n \cdot (n + 1) \tag{2.1}$$

◄

Incidentally, the same idea of doubling and then halving was also used by the 8-year-old Carl Friedrich Gauss when he solved his teacher's task in only a short time:

**Task: Add the (Natural) Numbers from 1 to 100**
To do this, write down the sum twice, the second time write down the summands in reverse order. Then 100 times two numbers are written one above the other, which together makes 101.
In total, twice the amount has the value $100 \cdot 101$, so the simple sum has the value $\frac{1}{2} \cdot 100 \cdot 101$, that is the value 5050.

$$
\begin{array}{ccccccccc}
1 & + & 2 & + & 3 & + \ldots + & 98 & + & 99 & + & 100 \\
100 & + & 99 & + & 98 & + \ldots + & 3 & + & 2 & + & 1 \\
\hline
101 & + & 101 & + & 101 & + \ldots + & 101 & + & 101 & + & 101
\end{array}
$$

**Alternative Derivation of the Sum Formula**
Instead of the rectangular form, you can also use a square arrangement of the stones to derive the sum formula.

**Example: Determine the Sum $1+2+3+\ldots+10$**
The square with the side length 11 contains $11^2 = 121$ stones, of which 11 stones lie in the diagonal, so half of $121 - 11 = 110$, that is, 55 stones below and above the diagonal.

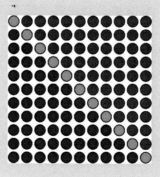

**General Derivation**
The respective $1 + 2 + \ldots + n$ blue and red stones together with the $n$ green diagonal stones form a square of the side length $(n + 1)$. So the following applies:

$$2 \cdot (1 + 2 + \ldots + n) + (n + 1) = (n + 1)^2$$

Transformation of the equation leads to:

$$2 \cdot (1 + 2 + \ldots + n) = (n + 1)^2 - (n + 1)$$

On the right side of the equation you can exclude the common term $(n + 1)$:

$2 \cdot (1 + 2 + \ldots + n) = (n + 1) \cdot [n + 1 - 1]$, so:

$$1 + 2 + \ldots + n = \tfrac{1}{2} \cdot n \cdot (n + 1)$$

**Suggestions for Reflection and for Investigations**

**A 2.1:** Determine systematically the sum of the first $n$ natural numbers with greatest summands $n = 1, 2, 3 \ldots, 20$.

**A 2.2:** Up to which natural number $n$ has one to sum up, until the sum 100 [1000; 1,000,000] is exceeded?

Another (third) way of deriving the formula results from the following square arrangement of stones: This is done by combining rows of 1 to 10 blue stones with rows of 1 to 9 red stones.

It is common to name the sequence of the subtotals of the first $n$ natural numbers, that is, the sequence of numbers 1, 3, 6, 10, 15, … as the sequence of the **triangular numbers** $(\Delta_n)_{n \in \mathbb{N}}$ (see also Chap. 16).

The $n$-th triangular number $\Delta_n$ can, therefore, be represented as follows:

$$\Delta_n = \tfrac{1}{2} \cdot n \cdot (n + 1)$$

Thus the figure illustrates the sum of the 10th triangular number and the 9th triangular number – together they form a square of the side length 10. So the following is valid:

$$\Delta_9 + \Delta_{10} = 10^2$$

Such combinations are possible for any triangular numbers and their predecessors:

$$\Delta_1 + \Delta_2 = 2^2 = 4; \ \Delta_2 + \Delta_3 = 3^2 = 9; \ \Delta_3 + \Delta_4 = 4^2 = 16; \ldots$$

---

**Rule**

**Sum of Two Consecutive Triangular Numbers**

The sum of two consecutive triangular numbers $\Delta_{n-1}$ and $\Delta_n$ is always a square number.

The following applies:

$$\Delta_{n-1} + \Delta_n = n^2 \tag{2.2}$$

◄

---

This property can be used as follows to derive the closed-form expression:

If one adds *n* to both sides of the equation.

$$(1 + 2 + 3 + \ldots + n) + (1 + 2 + 3 + \ldots + n - 1) = n^2,$$

then the following results

$$(1 + 2 + 3 + \ldots + n) + (1 + 2 + 3 + \ldots + n - 1) + n = n^2 + n, \text{ that is}$$
$$2 \cdot (1 + 2 + 3 + \ldots + n) = n \cdot (n + 1), \text{ and therefore}$$
$$1 + 2 + \ldots + n = \tfrac{1}{2} \cdot n \cdot (n + 1).$$

**Suggestions for Reflection and for Investigations**

**A 2.3:** The Greek mathematician Diophantus (about 250 A.D.) discovered a relationship between triangular and square numbers, which can be seen in the following figures. Which one is this? Draw corresponding figures also for a smaller number of dots.

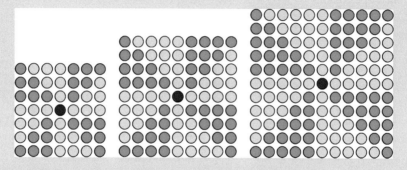

**A 2.4:** The following figures show dot patterns with the common property: number of red stones + number of yellow stones = number of orange stones

- Please explain: This property only applies if the side length $r$ of the square with the red stones and the side length $g$ of the square with the yellow stones are consecutive triangular numbers.
- This property is the reason why the following game of drawing from a hat is a fair game:

  In a hat there are $r$ red and $g$ yellow balls. Two balls are drawn one after the other and not put back. You win the game if the two drawn balls have the same color. (So you lose the game if the two drawn balls have different colors).

## 2.2   The Sum of the First $n$ Odd Natural Numbers

Square patterns can also be laid out in another way using colored stones: Around an existing square of $m \cdot m$ stones you place an L-shaped form, a so-called **gnomon,** with stones of a different color. For this purpose $m + 1 + m$, or $2m + 1$ stones are required, that is, an odd number of stones.

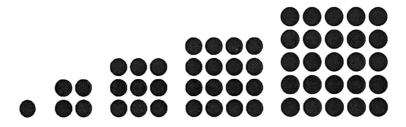

Around a "square" of $1 \cdot 1$ stones you can place an L-shaped form with $2 \cdot 1 + 1 = 3$ stones, around a square of $2 \cdot 2$ stones you can place an L-shaped form with $2 \cdot 2 + 1 = 5$ stones, etc.

So the following applies: $1 + 3 = 2^2$, $1 + 3 + 5 = 3^2$, $1 + 3 + 5 + 7 = 4^2$, ...

This can be generalized for any sum of the first $n$ odd natural numbers:

---

**Rule**

**Sum of the First $n$ Odd Numbers**

The sequence of the first $n$ odd natural numbers can be represented as a square pattern of L-shapes.

The sum of the first $n$ odd natural numbers is equal to the square number $n^2$.

$$1 + 3 + 5 + \ldots + (2n - 1) = n^2 \tag{2.3}$$

◀

**A Second Way of Deriving the Formula**

For the derivation, the method of doubling and then halving can be used again.

---

**Example: Sum of the Odd Natural Numbers from 1 to 9**

One after the other 9, then 7, then 5, then 3 blue stones, and finally 1 blue stone are arranged in rows and left-aligned on top of each other. In the second step, the rows are then filled up with red stones (1, 3, 5, 7, 9) to form a rectangle of 5 rows with 10 stones each.

Twice the amount $1 + 3 + 5 + 7 + 9$ is, therefore, equal to $5 \cdot 10 = 50$.

So for the sum $1 + 3 + 5 + 7 + 9$ you get $\frac{1}{2} \cdot 50 = 25$.

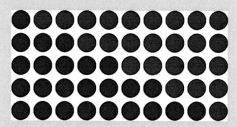

---

**General Derivation:**

$$2 \cdot (1 + 3 + 5 + \ldots + (2n - 1)) = n \cdot [(2n - 1) + 1], \text{ that is:}$$
$$1 + 3 + 5 + \ldots + (2n - 1) = \frac{1}{2} \cdot n \cdot 2n = n \cdot n = n^2$$

**A Third Way to Deduce the Formula**

Since the number of stones is odd, you can also arrange them in an axially symmetricical form.

**Example: Sum of the Odd Natural Numbers from 1 to 9**

The arrangement of the stones in an symmetrical form can be analyzed in different ways:

If you use three different colors for the stones, one has on the left and right side of the symmetry axis $1 + 2 + 3 + 4 = 10$ stones, additionally – since 5 odd numbers are added – the 5 stones of the symmetry axis.

After applying the sum formula, this results in

$$1 + 3 + 5 + 7 + 9 = 2 \cdot (1 + 2 + 3 + 4) + 5$$
$$= 2 \cdot \left( \tfrac{1}{2} \cdot 4 \cdot 5 \right) + 5$$
$$= 4 \cdot 5 + 5 = 5 \cdot 5 = 5^2$$

You can also see that a square results if you turn the triangle lying to the right of the axis by 180° and move it to the left.

**General Derivation:**

$$1 + 3 + 5 + \ldots + (2n + 1) = 2 \cdot (1 + 2 + 3 + \ldots + n - 1) + n$$
$$= 2 \cdot \left[ \tfrac{1}{2} \cdot (n - 1) \cdot n \right] + n$$
$$= (n - 1) \cdot n + n = n \cdot n = n^2$$

**Suggestions for Reflection and for Investigations**

**A 2.5:** Please explain: Another way to deduce formula (2.3) results from a fourfold symmetrical arrangement of stones.

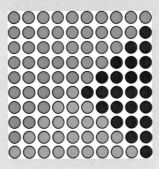

**A 2.6:** Find a general formula for the number of dots in the figures shown below

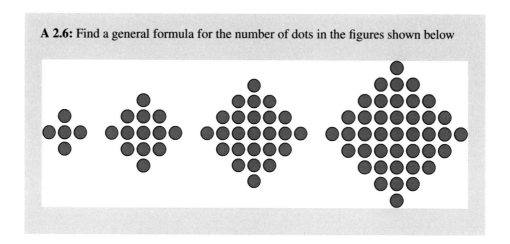

## 2.3    Quotients of Sums of Odd Natural Numbers

Galileo Galilei (1564–1642) noticed that for an even number of consecutive odd numbers a special property applies:

The sum of the first half of the odd numbers is always one third of the sum of the second half of the odd numbers, that is, if the quotient is formed, it is always equal to $\frac{1}{3}$:

$$\frac{1}{3} = \frac{1+3}{5+7} = \frac{1+3+5}{7+9+11} = \frac{1+3+5+7}{9+11+13+15} = \dots$$

So in general:

$$\frac{1+3+5+\dots+(2n-1)}{(2n+1)+(2n+3)+(2n+5)+\dots+(2n+(2n-1))} = \frac{1}{3}$$

However, this property is no longer surprising if one considers the symmetrical representation of odd natural numbers.

In the following figures the quotients can be taken from the ratio of the number of red to green/blue dots: In the first figure the ratio is 1:3, in the second figure the ratio is $(1+3):(5+7) = 4:12$, and in the third $(1+3+5):(7+9+11) = 9:27$.

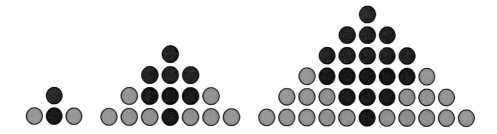

As another option one can consider the following illustration of odd numbers using the L-shaped forms.

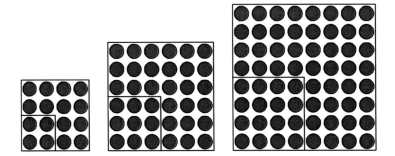

The fact that this property only applies to sums of odd natural numbers and not generally to sums of natural numbers can already be seen from the first elements of the sequence:

$$a_1 = \frac{1}{2} = 0.5; \; a_2 = \frac{1+2}{3+4} = \frac{3}{7} = 0.428571\ldots;$$

$$a_3 = \frac{1+2+3}{4+5+6} = \frac{6}{15} = 0.4; \; a_4 = \frac{1+2+3+4}{5+6+7+8} = \frac{10}{26} = 0.384615\ldots;$$

$$a_5 = \frac{1+2+3+4+5}{6+7+8+9+10} = \frac{15}{40} = 0.375;$$

$$a_6 = \frac{1+2+3+4+5+6}{7+8+9+10+11+12} = \frac{21}{57} = 0.368421\ldots$$

This is a strictly monotonically decreasing sequence with the limit $\frac{1}{3}$.

The limit can be read from the transformed sequence term:

$$a_n = \frac{\frac{1}{2} \cdot n \cdot (n+1)}{\frac{1}{2} \cdot 2n \cdot (2n+1) - \frac{1}{2} \cdot n \cdot (n+1)} = \frac{n^2 + n}{3n^2 + n} = \frac{n+1}{3n+1} = \frac{1+\frac{1}{n}}{3+\frac{1}{n}}$$

The fact that the sequence converges to the limit $\frac{1}{3}$ can also be illustrated graphically. In the following graphics the numerator sum is illustrated by red stones, the denominator sum by green, yellow, and light blue stones, whereby the triangle of yellow stones contains $n$ stones less than the others. With increasing $n$ this missing number of stones plays a smaller and smaller role.

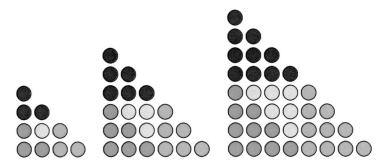

**Suggestions for Reflection and for Investigations**
**A 2.7** Explain: The last figures also illustrate a relationship between the $n$-th and $2n$-th triangular number, namely $\Delta_{2n} + n = 4 \cdot \Delta_n$.

## 2.4   Representation of a Natural Number as the Sum of Consecutive Natural Numbers

In Sect. 2.1 we explained how to calculate the sum of the first $n$ natural numbers. Using the term $\frac{1}{2} \cdot n \cdot (n + 1)$ one can also easily calculate sums of *any* consecutive natural numbers.

**Example: Sum of Natural Numbers Between 30 and 40 (incl.)**
You can get the sum of the 11 consecutive natural numbers between 30 and 40 by subtracting the sum of the first 29 natural numbers from the sum of the first 40 natural numbers:

$$\begin{aligned} 30 + 31 + \ldots + 40 &= (1 + 2 + \ldots + 40) - (1 + 2 + \ldots + 29) \\ &= \tfrac{1}{2} \cdot 40 \cdot 41 - \tfrac{1}{2} \cdot 29 \cdot 30 \\ &= 20 \cdot 41 - 29 \cdot 15 \\ &= 820 - 435 = 385 \end{aligned}$$

Generalization: You get the sum of all natural numbers between two natural numbers $a$ and $b$, by decreasing the natural numbers up to the greatest summand $b$ minus the sum of the natural numbers up to the predecessor of the smallest summand $a$, that is, $a - 1$.

**Example: Sum of Natural Numbers between 30 and 40 (incl.) – Alternative Method**

$$\begin{aligned} 30 + 31 + 32 + \ldots + 40 &= (30 + 0) + (30 + 1) + (30 + 2) + \ldots + (30 + 10) \\ &= 11 \cdot 30 + (0 + 1 + 2 + \ldots + 10) \\ &= 11 \cdot 30 + \tfrac{1}{2} \cdot 10 \cdot 11 \\ &= 11 \cdot \left(30 + \tfrac{1}{2} \cdot 10\right) = 385 \end{aligned}$$

Generalization: You get the sum of all natural numbers between two natural numbers $a$ and $b$ (incl.), by multiplying the amount of natural numbers which has to be summed (i.e., $b - a + 1$) by the sum of the smallest summand $a$ and half of the difference between the largest and smallest summand, that is, $\frac{1}{2} \cdot (b - a)$.

In this section, the following question will be examined: Which sum values can occur by adding any number of consecutive natural numbers?

Are there perhaps natural numbers that can *not* be represented as the sum of consecutive natural numbers?

An indirect answer to this question is given by a theorem which the English mathematician **James Joseph Sylvester** (1814–1897) discovered:

---

**Theorem**

**Number of Possible Representations as Sums of Natural Numbers (Sylvester's Theorem)**

The number of ways to represent a natural number $n$ as the sum of consecutive natural numbers is equal to the number of odd divisors ($>1$) of the number $n$. ◀

---

That this theorem applies will first be illustrated by the following examples:

---

**Example: Representation of the Number 70 as the Sum of Consecutive Natural Numbers**

The natural number 70 has the odd divisors 5, 7, and 35.

Because of $70 = 5 \cdot 14$ one can represent 70 in the form of a rectangle consisting of 5 columns with **14** stones each. Since 5 is an odd number, there is a central column (red).

The number of stones in the central column remains untransformed in the second representation. From the columns left of the center, 1 or 2 stones are removed and added symmetrically to the columns on the right (highlighted in yellow in the figure).

This creates a pattern of $12 + 13 + \mathbf{14} + 15 + 16 = 70$ stones, that is, the number 70 can be represented as the sum of 5 consecutive natural numbers.

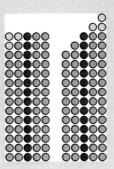

The same procedure can be applied to the second odd divisor, that is, the number 7:

The rectangle of $7 \times 10$ stones, that is, 7 columns, each with **10** dots, has a central column (red). Then again symmetrically to the central column 1 or 2 or 3 stones are taken away from the columns on the left and added to the columns on the right.

Thus we get

$$(10 - 3) + (10 - 2) + (10 - 1) + \mathbf{10} + (10 + 1) + (10 + 2) + (10 + 3) = 70, \text{ also}$$

$$7 + 8 + 9 + \mathbf{10} + 11 + 12 + 13 = 70.$$

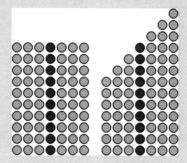

For the third odd divisor, the number 35, the previous procedure must be modified.

Although you can form a rectangle with $35 \times 2$ stones, and you can also remove 1 or 2 stones from the two adjacent columns left of the center, but from that point on, the number of stones would become "negative".

This can also be illustrated: We represent this in the form of stones that are *below* the previous drawing level.

So we have the following situation: Left of the central column (red), which consists of 2 stones, there is a column of 1 stone, then a column with 0 stones, then 15 columns with 1, 2, 3, ..., 15 stones in the negative range (yellow colored). To the right of the center there are columns with 3, 4, 5, ..., 20 stones.

The $1 + 2 + 3 + \ldots + 15$ stones in the negative range (yellow colored) are compensated by $1 + 2 + 3 + \ldots + 15$ stones in the positive range (colored red or green). What remains is a trapezoid of 4 columns with $16 + 17 + 18 + 19 = 70$ stones (orange colored) (Fig. 2.1).

**Summary**

The three possible representations of the number 70 as the sum of consecutive natural numbers are

$$12 + 13 + 14 + 15 + 16 = 70$$
$$\text{and}$$
$$7 + 8 + 9 + 10 + 11 + 12 + 13 = 70$$
$$\text{and}$$
$$16 + 17 + 18 + 19 = 70$$

**Fig. 2.1** Illustration of the case
with negative numbers

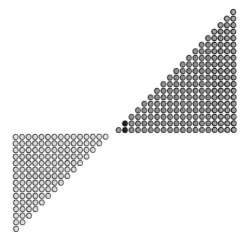

Obviously, a sum of consecutive natural numbers, consisting of an *odd* number of summands, represented as a staircase with an *odd* number of columns, can be transformed back to the rectangular form, that is, the natural number can be represented as the product of an *odd* number and the medial number of summands (the latter can be even or odd).

On the other hand, if one systematically explores the sums of an *even* number of consecutive natural numbers (each starting from the smallest summand 1), then it can be seen that they are all multiples of *odd* numbers, which means that here too the *odd* divisor of the natural number (the sum) plays a crucial role.

**Overview**
The odd numbers ($\geq 3$) themselves can be represented as the sum of *two* consecutive numbers: $1 \cdot 3 = 3 = 1 + 2$ and further $1 \cdot 5 = 5 = 2 + 3$; $1 \cdot 7 = 7 = 3 + 4$; $1 \cdot 9 = 9 = 4 + 5$ (see the following figures). The one row with 5, resp. 7, resp. 9 stones is not shown, but the modified arrangement with stones in the negative range.

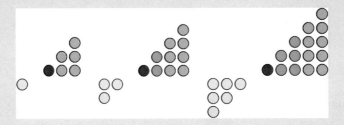

If you consider twice an odd number ($\geq 5$), this can be represented as the sum of *four* consecutive numbers: $2 \cdot 5 = 1 + 2 + 3 + 4$ and further $2 \cdot 7 = 2 + 3 + 4 + 5$; $2 \cdot 9 = 3 + 4 + 5 + 6$; $2 \cdot 11 = 4 + 5 + 6 + 7$ (see the following figures). Here the two rows with $2 \cdot 7$ or $2 \cdot 9$ or $2 \cdot 11$ stones are not shown, but the modified arrangement with stones in the negative range.

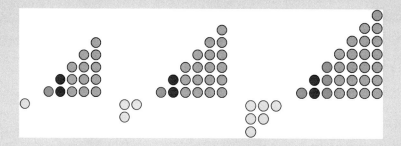

If you consider three times an odd number ($\geq 7$), this can be represented as the sum of *six* consecutive numbers: $3 \cdot 7 = 1 + 2 + 3 + 4 + 5 + 6$ and further $3 \cdot 9 = 2 + 3 + 4 + 5 + 6 + 7$; $3 \cdot 11 = 3 + 4 + 5 + 6 + 7 + 8$; $3 \cdot 13 = 4 + 5 + 6 + 7 + 8 + 9$ (see the following figures).

Here too, the three rows with $3 \cdot 9$ and $3 \cdot 11$ and $3 \cdot 13$ stones are not shown, however the modified arrangement with stones in the negative range.

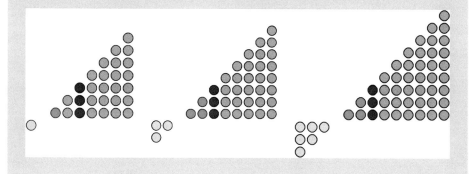

The procedure explained by means of examples can be applied to any natural number $n$:
If a natural number $n$ can be divided by an odd number $u$, then you can represent the number $n$ by a rectangle with the side lengths $u$ (width) and $x := \frac{n}{u}$ (height). This, therefore, consists of $u$ columns, each with $x$ stones. Since the number of columns is odd, the height of the central column can be maintained and that of the adjacent columns can be increased or decreased symmetrically by 1, 2, …

The rectangle of $u$ columns with $x$ stones each can be rearranged into a staircase figure, with $\frac{1}{2} \cdot (u - 1)$ columns each left and right to the central column which remains unchanged. The leftmost column then consists of $x - \frac{1}{2} \cdot (u - 1)$ stones, the rightmost column consists of $x + \frac{1}{2} \cdot (u - 1)$ stones.

If an odd divisor $u$ of a natural number $n$ is greater than $x=n/a$, then the matter becomes a little more complicated: the number $x - \frac{1}{2} \cdot (u - 1)$ of the stones in the leftmost column would be negative according to previous considerations. Columns in the negative range are balanced by columns of the same height which are in the positive range. The rightmost column which is needed for balancing, therefore, consists of $(-1) \cdot \left(x - \frac{1}{2} \cdot (u - 1)\right) = \frac{1}{2} \cdot (u - 1) - x$ stones.

So $u - 2 \cdot \left[\frac{1}{2} \cdot (u - 1) - x\right] - 1$ columns of a total of $u$ columns remain, that is, $2 \cdot x$ columns. This is always an even number.

---

**Theorem**

**Corollary to Sylvester's Theorem**

Of all natural numbers, only the powers of two (i.e., numbers of the form $n = 2^k$ with $k \in \mathbb{N}$) cannot be represented as a sum of consecutive natural numbers, because only these do not have odd divisors. ◄

---

**Suggestions for Reflection and for Investigations**

**A 2.8:** Examine in how many ways the natural numbers 18, 15, 45 can be represented as the sum of consecutive natural numbers. Visualize each of these possibilities using suitable patterns of colored stones.

**A 2.9:** Create an overview of the number of odd divisors for the natural numbers from 3 to 100.

**A 2.10:** The following applies $2020 = 2^2 \times 5 \times 101$. Which sum representations of consecutive natural numbers result for this year? Determine also the representations for the following years.

---

## 2.5    Sum of the First $n$ Square Numbers of Natural Numbers

The mathematician and physicist Abu Ali al-Hasan ibn al-Haitham (965–1039) is also known in Europe as Alhazen. He became particularly famous for his optical experiments, among other things he is considered as the "inventor" of the magnifying glass. Because of his numerous important discoveries he is also called the *father of optics*.

al-Haitham's ingenious approach to determine the sum of the first $n$ square numbers of natural numbers can even be generalized (although less descriptively) for sums of higher powers (see Chap. 16).

For the derivation of a formula only the knowledge of formula **(2.1)** for the sum of the first $n$ natural numbers is required, that is:

$$1 + 2 + 3 + \ldots + n = \tfrac{1}{2} \cdot n \cdot (n + 1)$$
$$= \tfrac{1}{2} \cdot n^2 + \tfrac{1}{2} \cdot n$$

His idea: You add suitable rectangles to the squares with the side lengths 1, 2, 3,..., $n$ – starting with a rectangle with height 1 up to a rectangle with height $n+1$.

**Example: Determine the Sum of the First Four Square Numbers**
The colored stones are arranged in the form of a rectangle, which has the width $1+2+3+4$ and the height $1+4$. In total, the squares contain $1^2, 2^2, 3^2$ and $4^2$ stones and the rectangular strips contain $1$, $1+2$, $1+2+3$ and $1, 1+2, 1+2+3$ and $1+2+3+4$ stones. It therefore applies:

$$\left(1^2 + 2^2 + 3^2 + 4^2\right) + (1) + (1+2) + (1+2+3) + (1+2+3+4)$$
$$= (1+2+3+4) \cdot 5$$

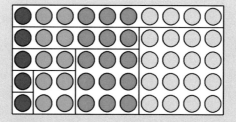

If the sums of natural numbers in brackets on the left side are replaced by the known term

$$1 = \tfrac{1}{2} \cdot 1^2 + \tfrac{1}{2} \cdot 1;$$
$$1 + 2 = \tfrac{1}{2} \cdot 2^2 + \tfrac{1}{2} \cdot 2;$$
$$1 + 2 + 3 = \tfrac{1}{2} \cdot 3^2 + \tfrac{1}{2} \cdot 3;$$
$$1 + 2 + 3 + 4 = \tfrac{1}{2} \cdot 4^2 + \tfrac{1}{2} \cdot 4,$$

it follows:

$$\left(1^2 + 2^2 + 3^2 + 4^2\right) + \left(\frac{1}{2} \cdot 1^2 + \frac{1}{2} \cdot 1\right) + \left(\frac{1}{2} \cdot 2^2 + \frac{1}{2} \cdot 2\right) +$$
$$\left(\frac{1}{2} \cdot 3^2 + \frac{1}{2} \cdot 3\right) + \left(\frac{1}{2} \cdot 4^2 + \frac{1}{2} \cdot 4\right) \quad = (1+2+3+4) \cdot 5$$

After removing the parentheses and rearranging, this is the next step:

$$\left(1^2 + 2^2 + 3^2 + 4^2\right) + \frac{1}{2} \cdot \left(1^2 + 2^2 + 3^2 + 4^2\right) + \frac{1}{2} \cdot (1 + 2 + 3 + 4)$$
$$= (1 + 2 + 3 + 4) \cdot 5$$

and further:

$$\tfrac{3}{2} \cdot \left(1^2 + 2^2 + 3^2 + 4^2\right) = \tfrac{9}{2} \cdot (1 + 2 + 3 + 4)$$

Solving this according for the sum of the square numbers we get

$$1^2 + 2^2 + 3^2 + 4^2 = \tfrac{9}{3} \cdot (1 + 2 + 3 + 4)$$

and after replacing the sum term for the first four natural numbers:

$$1^2 + 2^2 + 3^2 + 4^2 = \tfrac{9}{3} \cdot \tfrac{1}{2} \cdot (4^2 + 4) = 30$$

**Suggestions for Reflection and for Investigations**
**A 2.11:** Perform al-Haitham's approach for the sum of the first five square numbers.

### General Derivation of a Sum Formula for the First $n$ Square Numbers

All steps can be adapted accordingly for any $n \in \mathbb{N}$, that is, one generally considers a rectangle with the width $1 + 2 + 3 + \ldots + n$ and the height $n + 1$.

This is composed of $n$ squares with the area $1^2, 2^2, 3^2, \ldots, n^2$ and $n$ rectangular strips of height 1 with the area $1, 1 + 2, 1 + 2 + 3, \ldots, 1 + 2 + 3 + \ldots + n$. It therefore applies:

$$\left(1^2 + 2^2 + 3^2 + \ldots + n^2\right) + (1) + (1 + 2) + (1 + 2 + 3) + \ldots + (1 + 2 + 3 + \ldots + n)$$
$$= (1 + 2 + 3 + \ldots + n) \cdot (n + 1), \text{ that is}$$
$$\left(1^2 + 2^2 + 3^2 + \ldots + n^2\right) + \left(\tfrac{1}{2} \cdot 1^2 + \tfrac{1}{2} \cdot 1\right) + \left(\tfrac{1}{2} \cdot 2^2 + \tfrac{1}{2} \cdot 2\right) + \ldots + \left(\tfrac{1}{2} \cdot n^2 + \tfrac{1}{2} \cdot n\right)$$
$$= (1 + 2 + 3 + \ldots + n) \cdot (n + 1), \text{ that is}$$
$$\left(1^2 + 2^2 + 3^2 + \ldots + n^2\right) + \tfrac{1}{2} \cdot \left(1^2 + 2^2 + 3^2 + \ldots + n^2\right) + \tfrac{1}{2} \cdot (1 + 2 + 3 + \ldots + n)$$
$$= (1 + 2 + 3 + \ldots + n) \cdot (n + 1), \text{ that is}$$
$$\tfrac{3}{2} \cdot \left(1^2 + 2^2 + 3^2 + \ldots + n^2\right) = \left(n + \tfrac{1}{2}\right) \cdot (1 + 2 + 3 + \ldots + n).$$

Replacing the sum for the first $n$ natural numbers according to formula (2.1) results in

$$1^2 + 2^2 + 3^2 + \ldots + n^2 = \left(\tfrac{2}{3}n + \tfrac{1}{3}\right) \cdot (1 + 2 + 3 + \ldots + n), \text{ that is}$$
$$1^2 + 2^2 + 3^2 + \ldots + n^2 = \tfrac{1}{3} \cdot (2n + 1) \cdot \tfrac{1}{2} \cdot n \cdot (n + 1) = \tfrac{1}{6} \cdot n \cdot (n + 1) \cdot (2n + 1)$$

**Formula**

**Sum of the First $n$ Squares of Natural Numbers**

The sum of the first $n$ squares of natural numbers are represented by the formula

$$1^2 + 2^2 + 3^2 + \ldots + n^2 = \frac{1}{6} \cdot n \cdot (n+1) \cdot (2n+1) = \frac{1}{3}n^3 + \frac{1}{2}n^2 + \frac{1}{6}n \quad (2.4)$$

◄

**Suggestions for Reflection and for Investigations**

**A 2.12:** Set up a closed-form expression for the sum of the first $n$ odd square numbers.

**A 2.13:** The sum of two consecutive square numbers is an odd number and can, therefore, always be represented as the sum of two consecutive natural numbers, for example,

$$1^2 + 2^2 = 5 = 2 + 3; \ 2^2 + 3^2 = 13 = 6 + 7; \ 3^2 + 4^2 = 25 = 5^2 = 12 + 13$$

provide a general representation for the total sum $n^2 + (n+1)^2$.

**A 2.14:** How can one conclude from the following figure that the following applies: $1^2 + 2^2 + 3^2 + \ldots + n^2 = \frac{1}{6} \cdot n \cdot (n+1) \cdot (2n+1)$?

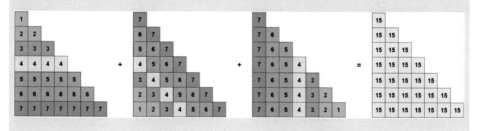

## 2.6 Sum of the First $n$ Cubes of the Natural Numbers

A formula for the sum of the first $n$ cubes was already known to the Greek mathematician Nicomachus of Gerasa (60–120 A.D.), as well as to the Indian mathematician Aryabhata (476–550 A.D.).

**Formula**

**Sum of the First $n$ Cubes of the Natural Numbers**

The sum of the first $n$ cubes of the natural numbers is equal to the square of the sum of the first $n$ natural numbers:

$$1^3 + 2^3 + 3^3 + \ldots + n^3 = \left(\frac{1}{2} \cdot n \cdot (n+1)\right)^2 = (1 + 2 + 3 + \ldots + n)^2 \quad (2.5)$$

◄

That this simple relationship applies has not been proved by these and other mathematicians of antiquity, because one could "see" this directly by comparison (see columns 3 and 5 of the following table).

| k | Σ k | (Σ k)² | k³ | Σ k³ |
|---|---|---|---|---|
| 1 | 1 | 1 | 1 | 1 |
| 2 | 3 | 9 | 8 | 9 |
| 3 | 6 | 36 | 27 | 36 |
| 4 | 10 | 100 | 64 | 100 |
| 5 | 15 | 225 | 125 | 225 |
| 6 | 21 | 441 | 216 | 441 |
| 7 | 28 | 784 | 343 | 784 |
| 8 | 36 | 1296 | 512 | 1296 |
| 9 | 45 | 2025 | 729 | 2025 |
| 10 | 55 | 3025 | 1000 | 3025 |

### 2.6.1 Proof of the Formula for the Sum of the First $n$ Cube Numbers by Al-Karaji

Abu Bakr ibn Muhammad ibn al-Husayn al-Karaji (953–1029), one of the most important mathematicians of the Islamic Middle Ages, was not satisfied with the *insight through mere looking,* but was the first to provide a formal (inductive) proof.

**Example: Proof of the Formula for n = 5**
A square with the side length $1 + 2 + 3 + 4 + 5$ can – according to a binomial formula – be divided into a square of the side length $1 + 2 + 3 + 4$ (blue–gray), in two rectangles of the length $1 + 2 + 3 + 4$ and width 5 (pink) and into a square of side length 5 (red).

$$(1 + 2 + 3 + 4 + 5)^2 = (1 + 2 + 3 + 4)^2 + 2 \cdot (1 + 2 + 3 + 4) \cdot 5 + 5^2$$

The sum of the first four natural numbers $1 + 2 + 3 + 4$ can be replaced by $\frac{1}{2} \cdot 4 \cdot 5$ so that it follows:

$$(1 + 2 + 3 + 4 + 5)^2 = (1 + 2 + 3 + 4)^2 + 2 \cdot \frac{1}{2} \cdot 4 \cdot 5^2 + 1 \cdot 5^2, \text{ that is}$$
$$(1 + 2 + 3 + 4 + 5)^2 = (1 + 2 + 3 + 4)^2 + 5 \cdot 5^2$$

i.e. the square with the side length $1 + 2 + 3 + 4 + 5$ can be dissected into a square of the side length $1 + 2 + 3 + 4$ and five squares of side length 5.

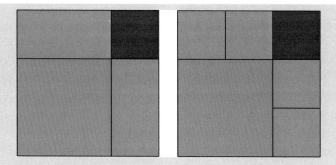

But this means that the following applies:

$$(1+2+3+4+5)^2 = (1+2+3+4)^2 + 5^3$$

Since this is also the case for the sums $(1+2+3+4)^2, (1+2+3)^2$, etc., follows step by step:

$$(1+2+3+4+5)^2 = (1+2+3+4)^2 + 5^3 = (1+2+3)^2 + 4^3 + 5^3$$
$$= (1+2)^2 + 3^3 + 4^3 + 5^3 = 1^2 + 2^3 + 3^3 + 4^3 + 5^3 = 1^3 + 2^3 + 3^3 + 4^3 + 5^3$$

That means, the square with the side length $1+2+3+4+5$ can be dissected into five squares of side length 5, four squares of side length 4, three squares of side length 3, two squares of side length 2, and one square of side length 1.

This can be seen in the following figure, where stones have been laid out in different colors (Fig. 2.2).

**Fig. 2.2** Representation of the sum of the first five cube numbers

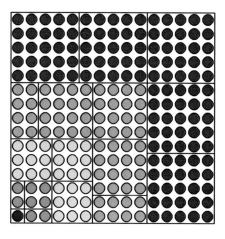

To see that this method of dissection of $1^3 + 2^3 + \ldots + n^3$ stones into a total of $n$ L-shapes also succeeds in the general case consider the following:

- Every $k$-th L-shaped form consists of $k$ subsquares with $k^2$ stones each $(k = 1, 2, \ldots, n)$ where for an even $k$ one of the subsquares of $k^2$ dots must be bisected.
- If $k$ is odd, $k^2$ stones lie in the "corner" square of the L-shaped form and in addition $\frac{1}{2} \cdot (k-1)$- times $k^2$ stones in horizontally or vertically arranged squares.
- If $k$ is even, $k^2$ stones are positioned at the "corner" square of the L-shaped form and in addition $\left(\frac{1}{2} \cdot k - 1\right)$-times $k^2$ stones are placed in horizontally or vertically arranged squares, additionally, half a square with $\frac{1}{2} \cdot k^2$ stones lies in the horizontally and the vertically part of the L-shaped form respectively.

If you look at the whole figure, then the width and height of this figure are determined by the $n$-th L-shaped form.

- If $n$ is an odd number, then the side of the whole square consists of $\frac{1}{2} \cdot (n-1) \cdot n + n = \frac{1}{2} \cdot n \cdot (n+1)$ stones, horizontally and vertically.
  The number $\frac{1}{2} \cdot n \cdot (n+1)$ implies according to the formula (2.1) nothing else than that the width and height of the whole square is equal to the sum $1 + 2 + \ldots + n$.
- If $n$ is even, then the side of the whole square consists of $\left(\frac{1}{2} \cdot n - 1\right) \cdot n + \frac{1}{2} \cdot n + n = \frac{1}{2} \cdot n \cdot (n+1)$ stones, horizontally and vertically.
  As in the odd case, this means that the width and height of the total square is equal to the sum $1 + 2 + \ldots + n$.

So it fits!

## 2.6.2   Proof of the Formula for Cube Numbers by Wheatstone

The English physicist and inventor Charles Wheatstone (1802–1875) described the relationship between the sum of the first $n$ cube numbers and the sum of the first $n$ natural numbers in the following way:

Cube numbers can be represented as the sum of consecutive odd numbers – more precisely:

The cube number $n^3$ can be represented as the sum of $n$ consecutive *odd* numbers.

The cube number $n^3 = n \cdot n^2$ can in fact be understood as $n$-times the sum of the square number $n^2$. And this sum in turn can be transformed so that a sum of consecutive odd numbers results.

**Example: Cube Numbers as the Sum of Consecutive Odd Natural Numbers**

$$1^3 = 1$$
$$2^3 = 2 \cdot 2^2 = 2 \cdot 4 = 4 + 4 = 3 + 5$$
$$3^3 = 3 \cdot 3^2 = 3 \cdot 9 = 9 + 9 + 9 = 7 + 9 + 11$$
$$4^3 = 4 \cdot 4^2 = 4 \cdot 16 = 16 + 16 + 16 + 16 = 13 + 15 + 17 + 19$$
$$5^3 = 5 \cdot 5^2 = 5 \cdot 25 = 25 + 25 + 25 + 25 + 25 = 21 + 23 + 25 + 27 + 29$$
$$6^3 = 6 \cdot 6^2 = 6 \cdot 36 = 36 + 36 + 36 + 36 + 36 + 36 = 31 + 33 + 35 + 37 + 39 + 41$$

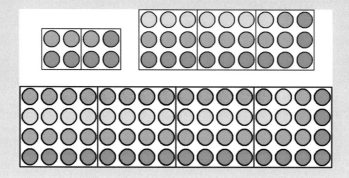

The figures consist – similar to the Sylvester figures – of $n$ stone columns (shown in the figure as lying columns). In the first $n-1$ squares the length of the columns is matching, but the last square contains exactly the same L-shaped forms as in the figures of Sect. 2.2, that is, the representation of the sum of the odd natural numbers $1 + 3 + 5 + \ldots + (2n - 1)$, see formula (2.3).

From the figures the following decomposition of the cube number $n^3$ is thus obtained:

$$n^3 = n \cdot n^2 = (n - 1) \cdot n^2 + 1 \cdot n^2 = (n - 1) \cdot n^2 + (1 + 3 + 5 + \ldots + (2n - 1))$$

- The shortest column, therefore, consists of $(n - 1) \cdot n + 1 = n^2 - n + 1$ stones,
- The next one is made of $(n - 1) \cdot n + 3 = n^2 - n + 3$ stones,
- The third one made of $(n - 1) \cdot n + 5 = n^2 - n + 5$ stones,
  and so on
- The longest column finally contains

$$(n - 1) \cdot n + (2n - 1) = n^2 - n + 2n - 1 = n^2 + n - 1 \text{ Stones.}$$

**Rule**

**Presentation of the Sum of the First $n$ Cube Numbers**

Each cube number $n^3$ can be represented as the sum of $n$ odd natural numbers whose mean value is $n^2$. The following applies,

- If $n$ is even:

$$n^3 = \left(n^2 - n + 1\right) + \ldots + \left(n^2 - 3\right) + \left(n^2 - 1\right) + \left(n^2 + 1\right) + \left(n^2 + 3\right) + \ldots + \left(n^2 + n - 1\right)$$

- If $n$ is odd:

$$n^3 = \left(n^2 - n + 1\right) + \ldots + \left(n^2 - 4\right) + \left(n^2 - 2\right) + n^2 + \left(n^2 + 2\right) + \left(n^2 + 4\right) + \ldots + \left(n^2 + n - 1\right)$$

According to formula (2.3) for the sum of the first $k$ odd numbers applies:

$$1 + 3 + 5 + \ldots + (2k - 1) = k^2$$

Here, one adds up to the odd number $\left(n^2 + n - 1\right)$. So if you put $2k - 1 = n^2 + n - 1$, it follows that $k = \frac{1}{2} \cdot \left(n^2 + n\right) = \frac{1}{2} \cdot n \cdot (n + 1)$.

If inserted, this results in

$$1^3 + 2^3 + 3^3 + \ldots + n^3$$
$$= 1 + [3 + 5] + [7 + 9 + 11] + \ldots + \left[\left(n^2 - n + 1\right) + \ldots + \left(n^2 + n - 1\right)\right]$$
$$= \left[\frac{1}{2} \cdot n \cdot (n + 1)\right]^2 = (1 + 2 + 3 + \ldots + n)^2$$

◄

---

**Suggestions for Reflection and for Investigations**
**A 2.15** When you add the first odd cubes, then a regularity is noticeable. You may discover a striking property:

- The sum of the first two odd cube numbers is equal to the sum of three consecutive powers of two (starting with exponent 2):

$$1^3 + 3^3 = 2^2 + 2^3 + 2^4 = 11100_2$$

- The sum of the first four odd cube numbers is equal to the sum of five consecutive powers of two (starting with exponent 4):

$$1^3 + 3^3 + 5^3 + 7^3 = 2^4 + 2^5 + 2^5 + 2^7 + 2^8 = 111110000_2$$

- The sum of the first eight odd cube numbers is equal to the sum of seven consecutive powers of two (starting with exponent 6):

$$1^3 + 3^3 + \ldots + 13^3 + 15^3 = 2^6 + 2^7 + \ldots + 2^{11} + 2^{12} = 1111111000000_2$$

- The sum of the first 16 odd cube numbers is equal to the sum of nine consecutive powers of two (starting with exponent 8):

$$1^3 + 3^3 + \ldots + 29^3 + 31^3 = 2^8 + 2^9 + \ldots + 2^{15} + 2^{16} = 11111111100000000_2$$

What are the "following" equations? Does this characteristic apply generally?

**A 2.16:** The sum of two consecutive cube numbers is an odd number and this can, therefore, always be represented as the sum of consecutive natural numbers, for example:

$$1^3 + 2^3 = 9 = 3^2 = 4 + 5 = 2 + 3 + 4;$$
$$2^3 + 3^3 = 35 = 17 + 18 = 5 + 6 + 7 + 8 + 9 = 2 + 3 + 4 + 5 + 6 + 7 + 8;$$
$$3^3 + 4^3 = 91 = 45 + 46 = 10 + 11 + 12 + 13 + 14 + 15 + 16$$
$$= 1 + 2 + 3 + 4 + 5 + 6 + 7 + 8 + 9 + 10 + 11 + 12 + 13.$$

Demonstrate that in general the sum of two cube numbers $n^3 + (n+1)^3$ can always be represented in at least two ways as the sum of consecutive natural numbers. Specify a general representation for these sums.

## 2.7 Pythagorean Triples

The arrangement of colored stones in the form of an L-shape can be used to find so-called **Pythagorean triples**. These are natural numbers $a, b, c$, which can be the side lengths of a right-angled triangle.

So we are looking for three suitable square numbers $a^2, b^2, c^2$, which satisfy the equation of the **Pythagorean theorem**

$a^2 + b^2 = c^2$ (see Chap. 17).

### 2.7.1 Simple Types of Pythagorean Triples

As is well known $(3, 4, 5)$ is a Pythagorean triple, because it is $3^2 + 4^2 = 5^2$.

The figure shows $3^2$ represented by the blue stones, $4^2$ by the red stones laid out in square form, and $5^2$ by the red stones laid out in a square form together with the blue stones in the form of an L-shape, which together also form a square pattern.

Since the number of stones of the (blue) L-shaped form in the example is a square number, the triple $(3, 4, 5)$ is a Pythagorean triple.

Basically, L-shaped forms that frame a square contain an odd number of stones. Further, Pythagorean triples can, therefore, be found by systematically considering the sequence of *odd square numbers*. The next odd square number after $3^2 = 9$ is $5^2 = 25$:

$5^2 = 25 = 13 + 12$ blue stones of the L-shaped form belong to a pattern of $12^2$ red stones and a total of $13^2$ (red and blue) stones (see the following figure).

In the following two figures, the largest of the three squares (the square on the hypotenuse) is shown, which belong to the next two odd square numbers and thus provides further Pythagorean triples:

- $7^2 = 49 = 25 + 24$ blue stones of the L-shaped form belong to a pattern of $24^2$ red stones and a total of $25^2$ (red and blue) stones.
- $9^2 = 81 = 41 + 40$ blue stones of the L-shaped form belong to a pattern of $40^2$ red stones and a total of $41^2$ (red and blue) stones.

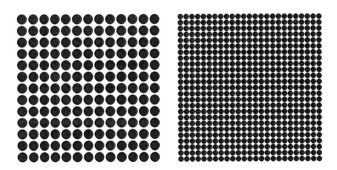

---

**Rule**

**Pythagorean Triples, for Whose Representation _a_ L-Shaped Form is Required**

In general one finds an infinite number of Pythagorean triples $(a_n;\ b_n;\ c_n)$ with $a_n = 2n + 1$, $b_n = 2n(n + 1)$, $c_n = 2n(n + 1) + 1$, i.e. $c_n - b_n = 1$. ◀

Proof: Starting from an odd square number (= number of blue stones of the L-shaped form), one calculates the number of red stones of the square which is framed by the blue L-shaped form: $(2n + 1)^2 = 4n^2 + 4n + 1$.

Considering the blue stones, *one* blue stone lies in the lower right corner of the square and half of the remaining blue stones lie below and to the right side of the red square respectively. The number $4n^2 + 4n + 1$ of the blue stones is, therefore, split as follows:

$$\text{Stone to the lower right} + \text{Stones in the lower row}$$
$$+ \text{Stone in the right column} = 1 + \left(2n^2 + 2n\right) + \left(2n^2 + 2n\right).$$

Because of $2n^2 + 2n = 2n \cdot (n + 1)$ the side length of the red square can, therefore, be specified with the term $b_n = 2n \cdot (n + 1)$. The whole square then has the side length $c_n = 2n \cdot (n + 1) + 1$.

The following table contains the first triples of this type:

| $n$ | $a_n = 2n + 1$ | $(2n + 1)^2$ | $b_n = 2n \cdot (n{+}1)$ | $c_n = 2n \cdot (n + 1) + 1$ |
|---|---|---|---|---|
| 1 | 3 | 9 = 4 + 5 | 4 | 5 |
| 2 | 5 | 25 = 12 + 13 | 12 | 13 |
| 3 | 7 | 49 = 24 + 25 | 24 | 25 |
| 4 | 9 | 81 = 40 + 41 | 40 | 41 |
| 5 | 11 | 121 = 60 + 61 | 60 | 61 |
| 6 | 13 | 169 = 84 + 85 | 84 | 85 |
| ... | ... | ... | ... | ... |

## 2.7.2 Further Pythagorean Triples

Apart from the infinite number of Pythagorean triples just described, there are others: One can just place *two* adjacent L-shaped forms, and if the number of stones in the two L-shaped forms *together* results in a square number, actually another Pythagorean triple has been found.

As *one* L-shaped form always contains an odd number of stones, a double L-shaped form always contains an even number of stones.

Therefore, you have to go through the sequence of the even square numbers, distribute this number of blue stones to the two L-shaped forms and from this determine the side length of the square of red stones.

The smallest possible even square number is $4^2 = 16$, which can be represented as the sum of the two consecutive odd numbers 7 and 9. The side length 3 of the red square results from the number $7 = 3 + 1 + 3$ of the inner L-shaped form.

The first triple of numbers is (3, 4, 5) with $7 + 9 = 16 = 4^2 = (3 + 4) + (4 + 5)$ blue stones, $3^2$ red stones, and a total $5^2$ red and blue stones. We have already found this when exploring the type with *one* L-shaped form, but as will be shown in the following, here also an infinite number of *different* Pythagorean triples exists.

Analogously, the next even square numbers are found:

- $6^2 = 36 = 17 + 19 = (8 + 9) + (9 + 10)$ blue stones of the two L-shaped forms belong to a pattern of $8^2$ red stones and a total of $10^2$ (red and blue) stones,
- $8^2 = 64 = 31 + 33 = (15 + 16) + (16 + 17)$ blue stones of the two L-shaped forms belong to a pattern of $15^2$ red stones and a total of $17^2$ stones (see the following illustrations).

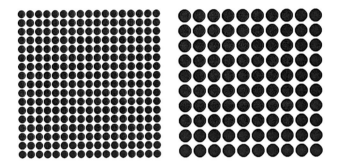

The next triple of this type then results as follows:

$10^2 = 100 = 49 + 51 = (24 + 25) + (25 + 26)$ blue stones of the two L-shaped forms belong to a pattern of $24^2$ red stones and a total of $26^2$ stones, etc.

---

**Rule**

**Pythagorean Triples, for Whose Representation *Two* L-Shaped Forms are Required**

In general one finds an infinite number of Pythagorean triples $(a_n; b_n; c_n)$ with $a_n = 2n$, $b_n = n^2 - 1$, $c_n = n^2 + 1$, i.e. $c_n - b_n = 2$. ◄

For the number of blue stones applies:

$$\underset{\text{even square}}{(2n)^2 = 4n^2} \quad = \underset{\text{inner L-Shape}}{\left[2n^2 - 1\right]} \quad + \underset{\text{outer L-Shape}}{\left[2n^2 + 1\right]}$$

The inner of the two L-shaped forms consists of $(n^2 - 1)$ blue stones on the lower side (= width of the red square), 1 stone in the lower right corner and $(n^2 - 1)$ stones on the right side of the red square. The outer of the two L-shaped forms consists of $n^2$ blue stones below, 1 stone in the lower right corner and $n^2$ stones right of the red square. The square of blue and red stones has a side length of $(n^2 + 1)$.

The following table contains the first Pythagorean triples of this type:

| $n$ | $a_n = 2n$ | $(2n)^2$ | $b_n = n^2 - 1$ | $c_n = n^2 + 1$ |
|---|---|---|---|---|
| 2 | 4 | 16  = 7 + 9 = (3 + 4) + (4 + 5) | 3 | 5 |
| 3 | 6 | 36 = 17 + 19 = (8 + 9) + (9 + 10) | 8 | 10 |
| 4 | 8 | 64 = 31 + 33 = (15 + 16) + (16 + 17) | 15 | 17 |
| 5 | 10 | 100 = 49 + 51 = (24 + 25) + (25 + 26) | 24 | 26 |
| 6 | 12 | 144 = 71 + 73 = (35 + 36) + (36 + 37) | 35 | 37 |
| 7 | 14 | 196 = 97 + 99 = (48 + 49) + (49 + 50) | 48 | 50 |
| ... | ... | ... | ... | ... |

If $n$ is an odd number the method of the two L-shaped forms does not result in new Pythagorean triples, but only in the double of triples found with *one* L-shaped form. (Since for odd $n$ the numbers $2n, n^2 - 1$ and $n^2 + 1$ are even, i.e. divisible by 2).

For even $n$ you get three numbers whose greatest common divisor is 1: $\gcd(a_n; b_n; c_n) = 1$.

Such triples are called **primitive Pythagorean triples.**

Generally speaking, one can find all Pythagorean triples by the method of the L-shaped forms, because in the equation $a^2 + b^2 = c^2$ the symbol $b^2$ represents the stones of the red square and $c^2$ represents the stones of the whole square, and from this necessarily results $a^2$ as the number of stones of the L-shaped forms.

**Suggestions for Reflection and for Investigations**
**A 2.17:** If, analogous to the considerations with one or two L-shaped forms, one looks at a figure with three L-shaped forms, then one cannot find any new Pythagorean triples. Check this for the following diagram and give a general explanation.

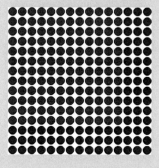

### 2.7.3 General Method for the Determination of all Pythagorean Triples

In literature you find a general method how to calculate *all* Pythagorean triples $(a; b; c)$ systematically. This method was already used by Euclid in his *Elements* (IX, prop. 28 and 29) and was probably even known to Babylonian mathematicians (about 1500 B.C.).

---

**Theorem**

**Representation of all Pythagorean Triples**

For any natural numbers $u$, $v$ with $u > v$ you get a Pythagorean triple $(a; b; c)$ using the approach $a = u^2 - v^2$, $b = 2uv$ and $c = u^2 + v^2$. ◄

---

This gives primitive number triples, if and only if $u$, $v$ are relatively prime and the sum of $u + v$ is an odd number.

Reasons for this approach are given at the end of this section.

If you systematically insert all natural numbers $u$, $v$ one after the other with $v < u$, you will get the left part of the Table 2.1. In the right part of the table, you can see how many L-shaped forms you would have needed to represent each triple, including the above examples.

Pythagorean triples were already found on Babylonian clay tablets from the time of the Hammurabi dynasty (1829–1530 B.C.). More than 3500 years ago, a calculation method must have been known – but certainly not the L-shaped form method, because the Pythagorean triple.

- (56, 90, 106) requires 16 or 50 L-shaped forms,
- (119, 120, 169) requires 49 or 50 L-shaped forms, and
- (12,709, 13,500, 18,541) requires 5041 or 5832 L-shaped forms,

see the following figures.

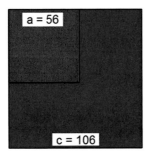
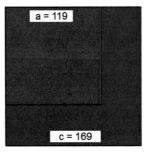

**Table 2.1**  List of Pythagorean triples and their representation by means of L-shaped forms

| $u$ | $v$ | $a = u^2 - v^2$ | $b = 2uv$ | $c = u^2 + v^2$ | Number of L-shaped forms |
|---|---|---|---|---|---|
| 2 | 1 | 3 | 4 | 5 | 1 or 2 |
| 3 | 1 | 8 | 6 | 10 | 2 or 4 |
| 3 | 2 | 5 | 12 | 13 | 1 or 8 |
| 4 | 1 | 15 | 8 | 17 | 2 or 9 |
| 4 | 2 | 12 | 16 | 20 | 4 or 8 |
| 4 | 3 | 7 | 24 | 25 | 1 or 18 |
| 5 | 1 | 24 | 10 | 26 | 2 or 16 |
| 5 | 2 | 21 | 20 | 29 | 8 or 9 |
| 5 | 3 | 16 | 30 | 34 | 4 or 18 |
| 5 | 4 | 9 | 40 | 41 | 1 or 32 |
| 6 | 1 | 35 | 12 | 37 | 2 or 25 |
| 6 | 2 | 32 | 24 | 40 | 8 or 16 |
| 6 | 3 | 27 | 36 | 45 | 9 or 18 |
| 6 | 4 | 20 | 48 | 52 | 4 or 32 |
| 6 | 5 | 11 | 60 | 61 | 1 or 50 |
| 7 | 1 | 48 | 14 | 50 | 2 or 36 |
| 7 | 2 | 45 | 28 | 53 | 8 or 25 |
| 7 | 3 | 40 | 42 | 58 | 16 or 18 |
| 7 | 4 | 31 | 56 | 65 | 9 or 34 |
| 7 | 5 | 24 | 70 | 74 | 4 or 50 |
| 7 | 6 | 13 | 84 | 85 | 1 or 72 |
| ... | ... | ... | ... | ... | ... |

## 2.7.4  Formula for Generating all Pythagorean Triples

The equation $x^2 + y^2 = z^2$ can be transformed into $\frac{x^2}{z^2} + \frac{y^2}{z^2} = 1$.

For relatively prime $x$, $z$ and $y$, $z$ a point on the unit circle with rational coordinates $a$, $b$ is described by $a = \frac{x}{z}$ and $b = \frac{y}{z}$.

For a straight line through the points $(-1, 0)$ and $(a, b)$ the equation applies:

$$y = \frac{b}{1 + a} \cdot (x + 1) + 0$$

This straight line intersects the $y$-axis in point $\left(0, \frac{b}{1+a}\right)$.

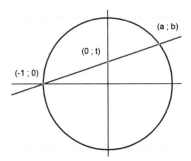

With $t = \frac{b}{1+a}$, you have $b = t \cdot (1 + a)$.

The points of the unit circle can, therefore, be determined by the equation

$$a^2 + t^2 \cdot (1 + a)^2 = 1$$

This equation can be transformed:

$$(a^2 - 1) + t^2 \cdot (1 + a)^2 = 0 \Leftrightarrow (a - 1)(a + 1) + t^2 \cdot (1 + a)^2 = 0 \Leftrightarrow (a + 1)\big[(a - 1) + t^2(a + 1)\big] = 0$$

Because of $0 \le a \le 1$, also $a + 1 > 0$, only the second factor can take on the value of zero, thus follows $(a - 1) + t^2(a + 1) = 0$ and further

$$a \cdot (1 + t^2) = 1 - t^2 \Leftrightarrow a = \frac{1 - t^2}{1 + t^2}.$$

This results in: $b = \frac{2t}{1+t^2}$

The points of the unit circle therefore can be determined as follows using only *one* parameter:

$$(a; b) = \left(\frac{x}{z}; \frac{y}{z}\right) = \left(\frac{1 - t^2}{1 + t^2}; \frac{2t}{1 + t^2}\right)$$

For a rational number $t$ with $0 < t < 1$, i.e. a fraction of the form $t = \frac{v}{u}$ with natural numbers $u$ as denominator and $v$ as numerator, where $v < u$, you get (after an intermediate step):

$$(a; b) = \left(\frac{1 - t^2}{1 + t^2}; \frac{2t}{1 + t^2}\right) = \left(\frac{u^2 - v^2}{u^2 + v^2}; \frac{2uv}{u^2 + v^2}\right)$$

Therefore, **all** Pythagorean triples $(x, y, z)$ can be obtained using the approach

$x = u^2 - v^2, y = 2uv, z = u^2 + v^2$ with $u > v$.

## 2.8     References to Further Literature

On **Wikipedia** you can find further information and literature with the keywords in English (German, French):

- Gaussian sum formula (Gaußsche Summenformel, -)
- Pythagorean triple (Pythagoreisches Tripel, Triplet pythagoric)

More informations can be found at **Wolfram Mathworld** with the keyword:

- Pythagorean triple

Many inspiring examples on the topic *Patterns of colored stones* can be found in Nelsen's books, see *general references.*

# Dissection of Rectangles into Largest Possible Squares

*Mathematics is a kind of toy that nature gave us to comfort and entertain us in the darkness.*

*(Jean-Baptiste le Rond d'Alembert, French mathematician, physicist and philosopher, 1717–1783)*

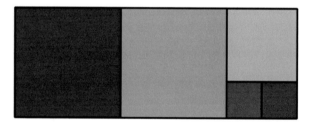

Can you determine the side lengths of the rectangle shown in the figure?

If the two small squares in the figure have the side length 1 unit (u), then for the light–blue-colored square the side length is 2 u, for the green and the blue square a side length of 3 u results. Therefore, the rectangle we started with has the side lengths 3 u and 8 u; we call it a $3 \times 8$ rectangle (where we name the "height" of the rectangle first).

If the smallest squares in the figure have the side length $a$ u, then the initial rectangle has the side lengths $3a$ u and $8a$ u.

## 3.1    A Game with a Rectangle

**Rule of the Game No. 1**

The starting player draws any rectangle on a sheet of paper ruled in squares. The sides of the rectangle lie on the lines of the grid. The second player then draws the largest

© Springer-Verlag GmbH Germany, part of Springer Nature 2021
H. K. Strick, *Mathematics is Beautiful*, https://doi.org/10.1007/978-3-662-62689-4_3

possible square flush with the left side of this rectangle, so that a smaller rectangle remains. Then it is again the first player's turn to draw a square as large as possible into the remaining rectangle, and so on.

The game ends when the initial rectangle is filled with squares.

The winner is the player who has drawn the last square.

**Rule of the Game No. 2**

(Modified game rule to speed up the game): When it is a player's turn, he draws the largest possible square into the rectangle (or the remaining rectangle). If it is possible to draw several squares of the same size into the remaining rectangle, then the same player may do so before it is the turn of the other player.

The game ends when the starting rectangle is filled with squares.

The winner is the player who was able to draw the last square.

---

**Example: Which Player Wins with a 3 × 8 Rectangle?**

In order to decide who would have won according to rule No. 1 or No. 2, the individual steps for dividing the $3 \times 8$ rectangle into largest possible squares must be examined:

In the first step a square with the side length 3 u can be drawn. This is colored blue in the figure.

The remaining rectangle then has the side lengths 5 u and 3 u. Into this fits maximally a square with the side length 3 u. This is colored green.

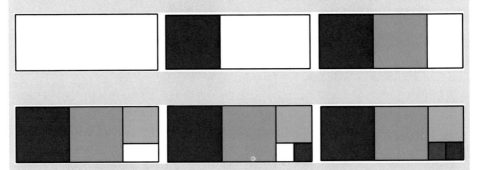

A rectangle with the side lengths 3 u and 2 u remains. You can draw maximally a square with the side length 2 u into this rectangle. This is colored light blue. A rectangle with the side lengths 2 u and 1 u remains, in which one can draw two squares with the side length 1 u. These are colored red and magenta.

The fact that two large squares with the side length 3 u, a medium square with the side length 2 u, and two small squares with the side length 1 u successively fit into the initial rectangle, becomes also clear by the following calculation:

$$3 \cdot 8 = 3 \cdot (3 + 5) = 3^2 + 3 \cdot 5$$

$$3 \cdot 5 = 3 \cdot (3 + 2) = 3^2 + 3 \cdot 2$$
$$2 \cdot 3 = 2 \cdot (2 + 1) = 2^2 + 2 \cdot 1$$
$$1 \cdot 2 = 1 \cdot (1 + 1) = 1^2 + 1^2$$

Summarization:

$$3 \cdot 8 = 2 \cdot 3^2 + 1 \cdot 2^2 + 2 \cdot 1^2$$

Since the first player drew the rectangle and a total of 2+1+2, that is, 5 squares could be placed, the second player wins because he could draw the last square.

The second player would also win according to rule No. 2, because three different squares were drawn into the rectangle, so the second player could draw the last two squares.

**Suggestions for Reflection and for Investigations**

**A 3.1:** Is it possible that the starting player chooses the side lengths of the rectangle so that he wins the game according to rule No. 1 and to rule No. 2?

Are there rectangles with which the first player wins the game according to rule No. 1 but loses according to rule No. 2 or vice versa?

**By the way**

The game, which is popular with children of primary school age, is well suited to develop their numeracy skills. However, they are more likely to describe the process in words than in formal notation.

## 3.2 Mathematical Analysis of the Game—Description Using Continued Fractions

Is it possible to decide who has won the game from Sect. 3.1 without drawing it, but solely from the knowledge of the side lengths by a quick calculation?

The largest possible square that fits into a rectangle has a side length that corresponds to the smaller side length of the rectangle. If you want to know how many of these squares fit into the rectangle, you just have to check how often the smaller side length fits into the larger one, that is, you have to calculate the quotient *larger side length* divided by *smaller side length*.

In the remaining rectangle, the originally shorter side is now the longer side, and the shorter side of the rectangle now to be considered is obtained by subtracting the smaller side length from the larger one (possibly several times), and so on.

**Example: Mathematical Analysis of the Dissection of the 3 × 8 Rectangle**

For the 3 × 8 rectangle, the quotient 8:3 must therefore be considered first; the result is 2,

that is, *two* squares with side length 3 can be drawn.

Then the quotient 3:2 is considered; the result is 1,

that is, *one* square of side length 2 can be placed.

In the third step, it is then about the quotient 2:1,

that is, *two* squares of side length 1 are drawn.

Those who are familiar with fraction rules can also perform this calculation as follows:

$$\frac{8}{3} = 2 + \frac{2}{3} = 2 + \frac{1}{\frac{3}{2}} = \underline{2} + \cfrac{1}{\underline{1} + \frac{1}{\underline{2}}}$$

After the first step, the inverse (reciprocal) of the fraction $\frac{2}{3}$ must be considered to check which maximum square fits into the $3 \times 2$ rectangle – instead of the $3 \times 2$ rectangle, we therefore consider a rectangle rotated by $90°$, that is, the $2 \times 3$ rectangle.

A representation of mixed fractions (= natural number plus proper fraction), where the denominator again contains mixed fractions, is called a **continued fraction.** If the numerators of the continued fraction are all ones, then one speaks of a **regular continued fraction.**

The numbers underlined here in the term indicate how many large, medium, or small squares can be drawn, if one starts with a 8 × 3-rectangle. From the underlined numbers you can immediately see which player has won according to rule No. 1 or No. 2.

The underlined numbers define the geometric figure except for multiples of the side length of the smallest square. The initial rectangle can also be briefly characterized by the following notation: [2; 1, 2].

Calculation and notation will be explained in another example:

**Example: Dissection of a 5 × 12 Rectangle into Largest Possible Squares**

For a rectangle with the side lengths 5 u and 12 u the following steps result:

$$5 \cdot 12 = 5 \cdot (5 + 7) = 5^2 + 5 \cdot 7$$
$$5 \cdot 7 = 5 \cdot (5 + 2) = 5^2 + 5 \cdot 2$$
$$2 \cdot 5 = 2 \cdot (2 + 3) = 2^2 + 2 \cdot 3$$
$$2 \cdot 3 = 2 \cdot (2 + 1) = 2^2 + 2 \cdot 1$$
$$1 \cdot 2 = 1 \cdot (1 + 1) = 1^2 + 1^2$$

Summarized: $5 \cdot 12 = 2 \cdot 5^2 + 2 \cdot 2^2 + 2 \cdot 1^2$.

From the continued fraction expansion

$$\frac{12}{5} = 2 + \frac{2}{5} = 2 + \frac{1}{\frac{5}{2}} = 2 + \frac{1}{2 + \frac{1}{2}}$$

this results in the representation $[2; 2, 2]$.

**Suggestions for Reflection and for Investigations**

**A 3.2:** In each case, develop the continued fraction that belongs to the dissection of the following rectangle:

(1) $3 \times 10$  (2) $4 \times 13$  (3) $5 \times 11$  (4) $6 \times 17$  (5) $7 \times 16$  (6) $8 \times 19$

## 3.3 Relationship Between the Continued Fraction Expansion and Rectangles

The two examples considered so far make it clear that in reverse a rectangle can also be determined as a visualization of a continued fraction. Except for multiples, this assignment is unique.

The following examples will illustrate this.

**Examples: Continued Fractions and Their Corresponding Rectangles**

To the continued fraction $[1; 3, 2] = \underline{1} + \frac{1}{\underline{3}+\frac{1}{2}} = 1 + \frac{1}{\frac{7}{2}} = 1 + \frac{2}{7} = \frac{9}{7}$

belongs a $7 \times 9$ rectangle (explicit dissection: $7 \cdot 9 = \underline{1} \cdot 7^2 + \underline{3} \cdot 2^2 + \underline{2} \cdot 1^2$), see the first of the following figures.

To the continued fraction $[1; 1, 2] = \underline{1} + \frac{1}{\underline{1}+\frac{1}{2}} = 1 + \frac{1}{\frac{3}{2}} = 1 + \frac{2}{3} = \frac{5}{3}$

belongs a $3 \times 5$ rectangle (explicit dissection: $3 \cdot 5 = \underline{1} \cdot 3^2 + \underline{2} \cdot 2^2 + \underline{2} \cdot 1^2$), see the second of the following figures.

To the continued fraction $[1; 2, 3] = \underline{1} + \frac{1}{2 + \frac{1}{3}} = 1 + \frac{1}{\frac{7}{3}} = 1 + \frac{3}{7} = \frac{10}{7}$

belongs a $7 \times 10$ rectangle (explicit dissection: $7 \cdot 10 = \underline{1} \cdot 7^2 + \underline{2} \cdot 3^2 + \underline{3} \cdot 1^2$),
see the third of the following figures.

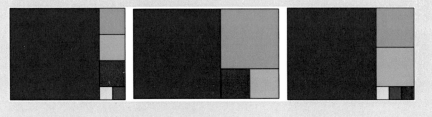

From the examples considered so far, it is clear that a rectangle can be created for any continued fraction $[a_0; a_1, a_2]$ using natural numbers $a_0, a_1, a_2$, provided that $a_2$ is at least 2.

To do this, you first draw $a_2$ squares with side length 1 u, over it then $a_1$ squares of the side length $a_2 \cdot 1$ u, that is $a_2$ u, and then $a_0$ squares with side length $a_2 \cdot a_1 + 1$. Therefore the rectangle has the side lengths $a_2 \cdot a_1 + 1$ and $a_0 \cdot (a_2 \cdot a_1 + 1) + a_2$:

---

**Rule**

**Rectangles Made up of Three Squares of Different Sizes**
The continued fraction $[a_0; a_1, a_2]$ can be assigned to a rectangle with the side lengths $a_0 \cdot a_1 \cdot a_2 + a_0 + a_2$ and $a_1 \cdot a_2 + 1$, which can be dissected into squares of three different sizes.

$$[a_0; a_1, a_2] = a_0 + \cfrac{1}{a_1 + \cfrac{1}{a_2}} = a_0 + \cfrac{1}{\frac{a_1 \cdot a_2 + 1}{a_2}} = a_0 + \cfrac{a_2}{a_1 \cdot a_2 + 1} = \cfrac{a_0 \cdot a_1 \cdot a_2 + a_0 + a_2}{a_1 \cdot a_2 + 1}$$

$$(3.1)$$

◀

---

**Suggestions for Reflection and for Investigations**
**A 3.3:** Determine the continued fraction expansion of $[a_0; a_1, a_2, a_3]$.
**A 3.4:** State the reason why the digit 1 cannot be placed at the last position of a continued fraction $[a_0; a_1, a_2, \cdots, a_n]$?
**A 3.5:** Which rectangle belongs to the continued fraction $[1; 2, 3, 4]$ (see the following figure)?

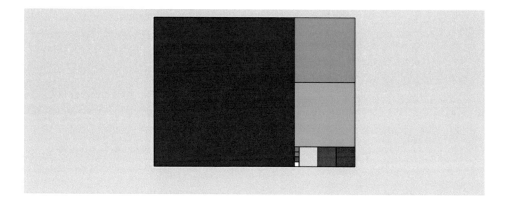

## 3.4   Dissection of Special Rectangles—Fibonacci Rectangles

A special role play those rectangles which are dissected into the largest possible squares and where only *one* square of each size is needed – except for the smallest square, which occurs twice. If the squares are arranged in form of an arc in a clockwise direction, then one can draw a quarter circle into each of the individual squares, so that altogether a spiral is created.

In Fig. 3.1a–d, first examples are shown. These are:

- A $5 \times 8$ rectangle with the associated continous fraction expansion:

$$\frac{8}{5} = 1 + \frac{3}{5} = 1 + \frac{1}{\frac{5}{3}} = 1 + \frac{1}{1 + \frac{2}{3}} = 1 + \frac{1}{1 + \frac{1}{\frac{3}{2}}} = 1 + \frac{1}{1 + \frac{1}{1 + \frac{1}{2}}} = [1; 1, 1, 2]$$

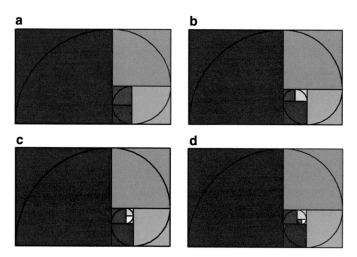

**Fig. 3.1   a–d** Dissection of special rectangles into largest possible squares

- A $8 \times 13$ rectangle with the associated continued fraction expansion:

$$\frac{13}{8} = 1 + \frac{5}{8} = 1 + \frac{1}{\frac{8}{5}} = 1 + \frac{1}{1+\frac{3}{5}} = 1 + \frac{1}{1+\frac{1}{\frac{5}{3}}} = \cdots = 1 + \frac{1}{1+\frac{1}{1+\frac{1}{1+\frac{1}{2}}}} = [1; 1, 1, 1, 2]$$

- A $13 \times 21$ rectangle with the associated continued fraction expansion:

$$\frac{21}{13} = 1 + \frac{8}{13} = 1 + \frac{1}{\frac{13}{8}} = 1 + \frac{1}{1+\frac{5}{8}} = 1 + \frac{1}{1+\frac{1}{\frac{8}{5}}} = \cdots = 1 + \frac{1}{1+\frac{1}{1+\frac{1}{1+\frac{1}{1+\frac{1}{2}}}}} = [1; 1, 1, 1, 1, 2]$$

- A $21 \times 34$ rectangle with the associated continued fraction expansion:

$$\frac{34}{21} = 1 + \frac{13}{21} = 1 + \frac{1}{\frac{21}{13}} = 1 + \frac{1}{1+\frac{8}{13}} = \cdots = 1 + \frac{1}{1+\frac{1}{1+\frac{1}{1+\frac{1}{1+\frac{1}{1+\frac{1}{2}}}}}} = [1; 1, 1, 1, 1, 1, 2]$$

The side lengths of the considered rectangles are adjacent elements of the sequence of the **Fibonacci numbers.** This is a sequence of numbers, whose elements are calculated by adding the two preceding elements: 1, 1, 2, 3, 5, 8, 13, 21, 34, 55, …

If you look closely in Fig. 3.1, you will see that in the next step, only *one* square is added to the rectangle just considered; thus a $5 \times 8$ rectangle becomes an $8 \times 13$ rectangle, an $8 \times 13$ rectangle becomes a $13 \times 21$ rectangle, and so on. Maybe you do not see this at first glance, because for the coloring of the rectangles in the figures blue is always used for the largest square, green for the next, and so on.

From all these figures, however, a regularity becomes clear:

---

**Theorem**

**Properties of Fibonacci Rectangles**

- A rectangle, whose side lengths are two adjacent elements $f_n$ and $f_{n+1}$ of the Fibonacci sequence can be dissected into a sequence of squares whose side lengths are exactly equal to the numbers of the Fibonacci sequence from $f_1$ to $f_n$.
- Into each of these squares quarter circles can be drawn, which altogether form a spiral curve.
- The sum of the squares of the first $n$ elements of the Fibonacci sequence $f_1^2 + f_2^2 + f_3^2 + \cdots + f_n^2$ (= total area of the $n$ squares) is equal to the product of the $n$-th Fibonacci number with the $(n+1)$-th number of the sequence, i.e.:

$$f_1^2 + f_2^2 + f_3^2 + \cdots + f_n^2 = f_n \cdot f_{n+1}$$

◀

---

**Suggestions for Reflection and for Investigations**

**A 3.6:** Draw the dissections of the rectangles whose side lengths are the first adjacent elements of the Fibonacci sequence, that is, the dissections of the $f_2 \times f_3$-rectangle, the $f_3 \times f_4$-rectangle, the $f_4 \times f_5$-rectangle.

## 3.5    The Sequence of Fibonacci Numbers

The elements of the Fibonacci sequence are defined recursively:

$$f_1 = 1; f_2 = 1 \text{ and } f_n = f_{n-1} + f_{n-2} \text{ for } n > 2,$$

so:

$$f_1 = 1; \ f_2 = 1; \ f_3 = f_1 + f_2 = 1 + 1 = 2$$
$$f_4 = f_2 + f_3 = 1 + 2 = 3; \ f_5 = f_3 + f_4 = 2 + 3 = 5$$
$$f_6 = f_4 + f_5 = 3 + 5 = 8; \ f_7 = f_5 + f_6 = 5 + 8 = 13$$

...

The French mathematicians Abraham de Moivre (1667–1754) and Jacques Philippe Marie Binet (1786–1856) discovered a method of calculating the Fibonacci numbers directly:

---

**Theorem**

**Moivre–Binet Formula**

The $n$-th element of the Fibonacci sequence (the $n$-th Fibonacci number) is calculated as follows

$$f_n = \frac{1}{\sqrt{5}} \cdot \left( \left( \frac{1 + \sqrt{5}}{2} \right)^n - \left( \frac{1 - \sqrt{5}}{2} \right)^n \right) \tag{3.2}$$

◀

In Europe, this numerical sequence became known through a book of the Italian mathematician Leonardo of Pisa in 1202, with the title *liber abaci* (freely translated: book of arithmetic). Leonardo's nickname was Fibonacci; it comes from Italian "figlio di Bonaccio", in English "Son of Bonaccio".

Probably the most famous problem of this book is the **the rabbit problem,** whose solution leads to the sequence of numbers that bears his name today:

*A newly born pair of rabbits becomes reproductive at the age of 1 month. After a gestation period of another month, the first offspring is born; it is again a pair of rabbits that reproduce in the same way as the first pair. How many pairs of rabbits then live in the n-th month?*

Assuming that the rabbits live forever and that the pairs of rabbits reproduce monthly (which reproduce again after 1 month), the following development of the population of rabbits can be observed:

After the 1st month only 1 pair (pair No. 1) exists; it now becomes reproductive, so that at the end of the 2nd month the first offspring pair (pair No. 2) is born, so 2 pairs exist.

After another month, pair No. 1 produces a pair of rabbits again (pair No. 3) – at the end of the 3rd month there are 3 pairs in total; now pair No. 2 is also capable of reproduction.

Pair No. 2 gives birth to a pair of rabbits at the end of the 4th month (pair No. 5), also pair No. 1; then there are 5 pairs alive, and so on (see also Fig. 3.2).

The numbers of the Fibonacci sequence play a role in numerous phenomena of nature (see references).

The stamps published in Macao/China in 2007 (see the following figure) take up the motif of the rabbit problem, show the arrangement of the seeds of sunflowers referring to

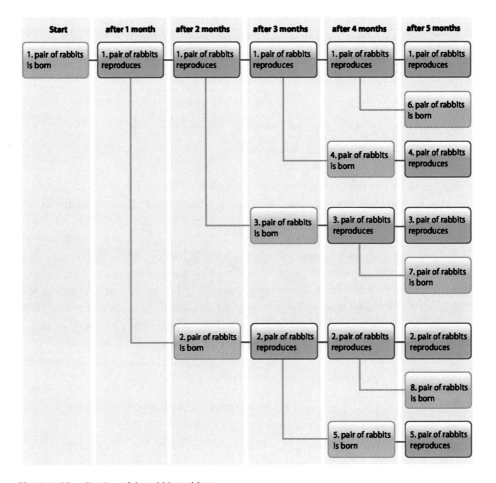

**Fig. 3.2** Visualization of the rabbit problem

the Penrose tiling (see also, Chap. 10) and show the section through a nautilus, a species of cephalopods found in the Indian Ocean, into a decomposed $13 \times 21$ rectangle.

The following stamps from Liechtenstein were issued in 2013 on the occasion of a mathematics exhibition with the title *Matheliebe (in English: Math Love)*, designed by Georg Schierscher. They show leaves of grapevine, a cultivated apple tree, and Japanese maple tree, whose structures are related to the "golden section." At the bottom edge of the stamps, you find the sequence of Fibonacci numbers (beginning with $f_2$, the sequence of digits of $f_{20} = 6765$ is truncated), the corresponding quotients (233:144 is truncated) and the **golden number** $\Phi = 1.61803\ldots$ (but with two errors: one digit 6 is missing at the 31st decimal place and the last digit should be 7).

The Swiss stamp was issued in 1987 on the occasion of the 150th anniversary of the Swiss Association of Engineers and Architects (Schweizerischen Ingenieur- und Architektenvereins=SIA) and shows the golden rectangle, where the side lengths correspond exactly to $\Phi$:1, with the first six squares of dissection together with quarter circles. The Spanish stamp was issued in 2006 on the occasion of the International Congress of Mathematicians in Madrid and is picking up the arrangement of the sunflower seeds as well.

## 3.6    Relationship with the Euclidean Algorithm

In his *Elements* Euclid explains a procedure for finding the greatest common divisor (gcd) of two given natural numbers $x$ and $y$.

He illustrates this by means of line segments whose lengths are $x$ and $y$. If two given numbers $x$, $y$ are interpreted as lengths of line segments $x = |AB|$ and $y = |CD|$, then the following applies:

*If CD does not measure AB, then, the less of the numbers AB, CD being continually subtracted from the greater, some number will be left which will measure the one before it.*

---

**Example: Application of the Euclidean Algorithm for Side Length of 28 u and 20 u**
We apply the Euclidean algorithm to determine the greatest common divisor of the numbers 20 and 28. To do this, we draw two line segments whose lengths are determined by the two numbers and reduce their length by mutual subtraction, see Table 3.1.

---

With the method of rectangle dissection described in the first sections, instead of drawing line segments that are subtracted, a rectangle is drawn with the initial numbers $x$, $y$ as side lengths (with e.g., $x > y$).

The largest possible square is cut off in this rectangle. This leaves a rectangle with the side lengths $y$ and $x - y$.

and so on.

**Table 3.1**   Steps to determine the gcd(20;28) according to Euclid

| | |
|---|---|
| | $y = 20$ does not exactly fit into $x = 28$; so you subtract 20 from 28 and it remains $x = 8$ and $y = 20$. |
| | $x = 8$ does not exactly fit into $y = 20$; so you subtract 8 from 20 and it remains $x = 8$ and $y = 12$. |
| | $x = 8$ does not exactly fit into $y = 12$; so you subtract 8 from 12 and it remains $x = 8$ and $y = 4$. |
| | $y = 4$ fits exactly into $x = 8$. |
| | Therefore the number 4 "measures" both 20 and 28. |

**Example: Application of the Euclidean Algorithm for a Rectangle with the Side Lengths 28 u and 20 u**

We look again at the natural numbers 20 and 28. To find the greatest common divisor of the two numbers, we draw a rectangle whose side lengths are determined by the two numbers and divide it into the largest possible squares, see Table 3.2.

**Table 3.2**   Steps to determine the gcd(20;28) using a rectangular dissection

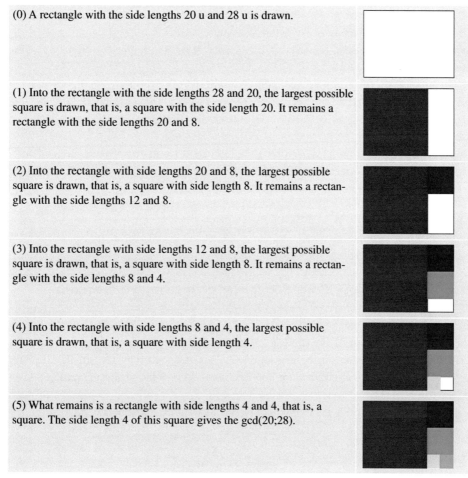

(0) A rectangle with the side lengths 20 u and 28 u is drawn.

(1) Into the rectangle with the side lengths 28 and 20, the largest possible square is drawn, that is, a square with the side length 20. It remains a rectangle with the side lengths 20 and 8.

(2) Into the rectangle with side lengths 20 and 8, the largest possible square is drawn, that is, a square with side length 8. It remains a rectangle with the side lengths 12 and 8.

(3) Into the rectangle with side lengths 12 and 8, the largest possible square is drawn, that is, a square with side length 8. It remains a rectangle with the side lengths 8 and 4.

(4) Into the rectangle with side lengths 8 and 4, the largest possible square is drawn, that is, a square with side length 4.

(5) What remains is a rectangle with side lengths 4 and 4, that is, a square. The side length 4 of this square gives the gcd(20;28).

---

**Rule**

**General Description of the Euclidean Algorithm**

In order to visualize the Euclidean algorithm, one needs only a few instructions:

1. Start: Define $x$, $y$.
2. If $x>y$ then draw a rectangle with the width $x$ and the height $y$, and a square with the side length $y$. Set $x:=x-y$. Go back to (2).
3. If $x<y$ then interchange $x$ and $y$ and go back to (2).
4. If $x=y$ then (finally) draw a square with the side length $x$. ◀

**Uniqueness of the Method of Drawing Rectangles**

With the drawing method, however, you can only determine the gcd if the side lengths are specified.

The figure for determining the gcd(20;28) looks exactly like the figure for the gcd(10;14) or for the gcd(5;7), but also for the gcd(15;21). The corresponding figures are *similar* to each other. For the smallest possible of the similar figures, the gcd of the side length is equal to 1.

Since the ratios of the side lengths are all equal, it results from the continued fraction expansion that the rectangular figure is composed of 1 large square, 2 medium-sized squares, and 2 small squares, in short: $[1; 2, 2]$:

$$\frac{28}{20} = \frac{21}{15} = \frac{14}{10} = \frac{7}{5} = 1 + \frac{2}{5} = 1 + \frac{1}{\frac{5}{2}} = \underline{1} + \frac{1}{\underline{2} + \frac{1}{\underline{2}}}$$

---

## 3.7    Examples of Infinite Sequences of Rectangle Dissections

If a natural number can be represented as the product of two coprime numbers $a$, $b$, then the representation of the product in the form of a rectangle with area $a \cdot b$ by the described method results in the dissection into squares, whereby the smallest squares have the side length 1 u; because for coprime numbers $a$, $b$ the greatest common divisor of the numbers is equal to 1.

Therefore, in the following, we will only consider coprime numbers $a$, $b$.

**Example: What do the Following Rectangles have in Common?**

$$a = 3,\ b = 5,\ a \cdot b = 15 = 3^2 + 2^2 + 1^2 + 1^2 \leftrightarrow [1; 1, 2] = \tfrac{5}{3} = 1.\overline{6}$$
$$a = 4, b = 7, a \cdot b = 28 = 4^2 + 3^2 + 1^2 + 1^2 + 1^2 \leftrightarrow [1; 1, 3] = \tfrac{7}{4} = 1.75$$
$$a = 5, b = 9, a \cdot b = 45 = 5^2 + 4^2 + 1^2 + 1^2 + 1^2 + 1^2 \leftrightarrow [1; 1, 4] = \tfrac{9}{5} = 1.8$$

The continued fraction is of the form $[1; 1, n]$ with $n = 2, 3, 4$.

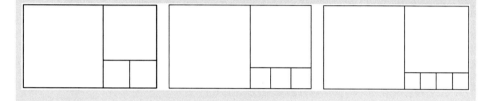

### Sequence of Continued Fractions with Limit 2

Which property can be discovered for continued fractions of the form $[1; 1, n]$ when the number $n$ grows unlimited?

Applying formula (3.1) results in:

$$[1; 1, n] = \frac{1 \cdot 1 \cdot n + 1 + n}{1 \cdot n + 1} = \frac{2n + 1}{n + 1}$$

For increasing $n$ this sequence converges toward the limit 2:

$$\frac{2n + 1}{n + 1} = \frac{2 + \frac{1}{n}}{1 + \frac{1}{n}} \to \frac{2}{1}$$

This means that the ratio of the side lengths of the rectangles gets closer and closer to 2 – the rectangle approaches more and more a double square.

### Sequence of Continued Fractions with Limit 10/7

The following, somewhat more complicated example can be used in a similar way:

Using the relationship

$$[a_0; a_1, a_2, a_3] = \frac{a_0 \cdot a_1 \cdot a_2 \cdot a_3 + a_0 \cdot a_1 + a_0 \cdot a_3 + a_2 \cdot a_3 + 1}{a_1 \cdot a_2 \cdot a_3 + a_1 + a_3}$$

(see **A 3.3**) results:

$$[1; 2, 3, n] = \frac{1 \cdot 2 \cdot 3 \cdot n + 1 \cdot 2 + 1 \cdot n + 3 \cdot n + 1}{2 \cdot 3 \cdot n + 2 + n} = \frac{10n + 3}{7n + 2} = \frac{10 + \frac{3}{n}}{7 + \frac{2}{n}} \to \frac{10}{7}$$

This means that the ratio of the side lengths of the rectangles gets closer and closer to 10/7 – the side lengths of the rectangles approach more and more the ratio 10:7.

**Suggestions for Reflection and for Investigations**

**A 3.7:** Analyze the sequences of the continued fractions with $[1; 2, n]$ and $[1; 1, 2, n]$ and determine their limits.

### The Infinite Sequence of Fibonacci Rectangles

In Sect. 3.4, we examined some examples of rectangles whose side lengths are adjacent Fibonacci numbers.

The quotients of adjacent numbers in the Fibonacci sequence (see Table 3.3) are alternately larger and smaller, with differences becoming smaller and smaller. If we neglect the last digit we see that the sequence converges against

$$[1; \overline{1}] = [1; 1, 1, 1, \ldots]$$

For two consecutive quotients the following relationship applies:

If the first quotient is written as $\frac{a}{b}$ then for the next quotient $\frac{a+b}{a}$ applies.

For the limit value $\Phi = \frac{a}{b}$, therefore, the following equation applies:

$$\frac{a}{b} = \frac{a+b}{a} \Leftrightarrow \frac{a}{b} = 1 + \frac{b}{a}$$

Inserting $\Phi = \frac{a}{b}$ gives (after transformation) a quadratic equation:

$$\Phi = 1 + \frac{1}{\Phi} \Leftrightarrow \Phi^2 = \Phi + 1 \Leftrightarrow \Phi^2 - \Phi + \frac{1}{4} = \frac{5}{4} \Leftrightarrow \left(\Phi - \frac{1}{2}\right)^2 = \frac{5}{4}$$

The positive solution is

$$\Phi = \frac{\sqrt{5}+1}{2} = 1.6180\ldots$$

**Theorem**

### The Golden Rectangle

The rectangles of the sequence $[1; \overline{1}] = [1; 1, 1, 1, \ldots]$ converge toward a rectangle with the following ratio of side lengths $\Phi : 1 = \frac{1}{2} \cdot \left(\sqrt{5}+1\right) : 1$ whereas $\Phi = \frac{\sqrt{5}+1}{2} = 1.6180\ldots$

It is called the **golden rectangle**, $\Phi$ is called the **golden number**. ◀

**Table 3.3** Sequence of the quotients of the side lengths of Fibonacci rectangles

| Quotient of the side lengths | $\frac{3}{2} = 1.5$ | $\frac{5}{3} = 1.\overline{6}$ | $\frac{8}{5} = 1.6$ | $\frac{13}{8} = 1.625$ | $\frac{21}{13} = 1.61538\ldots$ | $\ldots$ |
|---|---|---|---|---|---|---|
| Decomposition | $[1; 2]$ | $[1; 1, 2]$ | $[1; 1, 1, 2]$ | $[1; 1, 1, 1, 2]$ | $[1; 1, 1, 1, 1, 2]$ | $\ldots$ |

**Suggestions for Reflection and for Investigations**

**A 3.8:** Research the sequence of the quotients $f_{n+2}/f_n$ of Fibonacci numbers.

**Infinite Sequences of Rectangles with an Irrational Ratio of Side Length**

An interesting observation can be made if you continuously append an equal digit to the term of a continued fraction. The process is illustrated in the following figures.

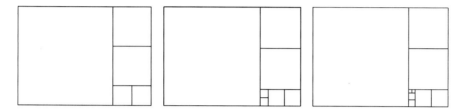

In this sequence, too, the quotient alternately takes smaller and larger values and also tends toward an assigned limit (see Table 3.4).

For two consecutive quotients the following correlation applies:

If the first quotient is written as $\frac{a}{b}$ then for the next quotient $\frac{a+2b}{a+b}$ applies. That is: you get the numerator of the next element of the sequence by adding the numerator to the double denominator, and the next denominator by adding numerator and denominator.

The limit must therefore apply:

$$\frac{a}{b} = \frac{a+2b}{a+b} \Leftrightarrow a \cdot (a+b) = (a+2b) \cdot b \Leftrightarrow a^2 + ab = ab + 2b^2 \Leftrightarrow a^2 = 2b^2 \Leftrightarrow \left(\frac{a}{b}\right)^2 = 2$$

The limit of the sequence is therefore equal to $[1; \overline{2}] = \sqrt{2} = 1.41421\ldots$

The rectangles in this sequence converge toward a rectangle with the ratio $\sqrt{2}{:}1$ (see the following figure).

*Note*: Such a rectangle can easily be constructed using the Pythagorean theorem.

**Table 3.4** Sequence of continued fractions with periodic last digit 2

| Ratio of the side lengths | $\frac{3}{2} = 1.5$ | $\frac{7}{5} = 1.4$ | $\frac{17}{12} = 1.41\overline{6}$ | $\frac{41}{29} = 1.4137\ldots$ | $\ldots$ |
|---|---|---|---|---|---|
| Decomposition | $[1;2]$ | $[1;2,2]$ | $[1;2,2,2]$ | $[1;2,2,2,2]$ | $\ldots$ |
| Continued fraction | $1 + \frac{1}{2}$ | $1 + \frac{1}{2+\frac{1}{2}}$ | $1 + \frac{1}{2+\frac{1}{2+\frac{1}{2}}}$ | $1 + \frac{1}{2+\frac{1}{2+\frac{1}{2+\frac{1}{2}}}}$ | $\ldots$ |

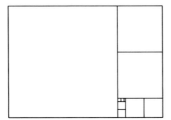

**Suggestions for Reflection and for Investigations**

**A 3.9:** Visualize the statement formulated above: "If the first quotient is denominated with $\frac{a}{b}$, then applies for the next quotient $\frac{a+2b}{a+b}$."

**A 3.10:** Determine the limit of the sequence of continued fractions $[1; 1, 2]$, $[1; 1, 2, 1, 2], [1; 1, 2, 1, 2, 1, 2], \ldots$

**A 3.11:** The elements of the sequence of continued fractions with limit $\sqrt{2}$ (see above) are good rational approximations for the irrational number. Explore the quality of approximation for the first elements of the sequence.

Proceed in the same way for the sequence of continued fractions from **A 3.10**.

## 3.8    Determination of Continued Fractions of Square Roots

To find the infinite continued fractions belonging to square roots of natural numbers, the same algebraic trick can be applied:

You consider the integer part of a root and the "remainder" separately, transform this term using the 3rd binomial formula, then replace the root in the denominator by the fraction that you obtained with the transformations—until the numerators of the fractions only contain ones …

Around 1760, Leonhard Euler proved the theorem that the continued fractions of the square roots of natural numbers are all periodic (except for square numbers).

**Expansion of $\sqrt{2}$ as a continued fraction**

$$\sqrt{2} = 1 + \left(\sqrt{2} - 1\right) = 1 + \frac{\left(\sqrt{2} - 1\right)\left(\sqrt{2} + 1\right)}{\left(\sqrt{2} + 1\right)} = 1 + \frac{1}{1 + \sqrt{2}} = 1 + \frac{1}{1 + \left(1 + \frac{1}{1+\sqrt{2}}\right)}$$

$$= 1 + \frac{1}{2 + \frac{1}{1 + \left(1 + \frac{1}{1+\sqrt{2}}\right)}} = 1 + \frac{1}{2 + \frac{1}{2 + \frac{1}{1+\sqrt{2}}}} = \cdots = \left[1; \overline{2}\right]$$

**Expansion of $\sqrt{3}$ as a continued fraction**

$$\sqrt{3} = 1 + \left(\sqrt{3} - 1\right) = 1 + \frac{\left(\sqrt{3} - 1\right) \cdot \left(\sqrt{3} + 1\right)}{\left(\sqrt{3} + 1\right)} = 1 + \frac{2}{1 + \sqrt{3}} = 1 + \frac{2}{2 + \frac{2}{1+\sqrt{3}}}$$

$$= 1 + \frac{1}{1 + \frac{1}{1+\sqrt{3}}} = 1 + \frac{1}{1 + \frac{1}{1+1+\frac{1}{1+\frac{1}{1+\sqrt{3}}}}} = 1 + \frac{1}{1 + \frac{1}{2 + \frac{1}{1+1+\frac{1}{1+\frac{1}{1+\sqrt{3}}}}}} = \cdots = \left[1; \overline{1, 2}\right]$$

First elements of the continued fraction expansion for $\sqrt{3}$ are illustrated in the following figure.

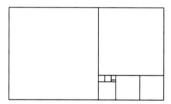

**Expansion of $\sqrt{5}$ as a continued fraction**

$$\sqrt{5} = 2 + \left(\sqrt{5} - 2\right) = 2 + \frac{\left(\sqrt{5} - 2\right) \cdot \left(\sqrt{5} + 2\right)}{\left(\sqrt{5} + 2\right)} = 2 + \frac{1}{2 + \sqrt{5}}$$

$$= 2 + \frac{1}{2 + \left(2 + \frac{1}{2+\sqrt{5}}\right)} = 2 + \frac{1}{4 + \frac{1}{2+\left(2+\frac{1}{2+\sqrt{5}}\right)}} = \cdots = \left[2; \overline{4}\right]$$

First elements of the continued fraction expansion for $\sqrt{5}$ are illustrated in the following figure.

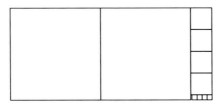

**Suggestions for Reflection and for Investigations**

**A 3.12:** Determine the continued fractions $\sqrt{6}$ and $\sqrt{7}$ (see also the following figures).

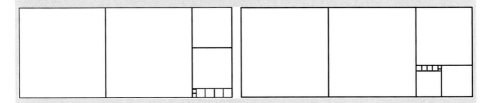

**A 3.13:** Which continued fractions result when the reciprocal of a continued fraction is calculated?

**A 3.14:** Calculate the continued fractions of quotients of consecutive square numbers. What properties do you guess?

## 3.9    References to Further Literature

On **Wikipedia** you can find further information and literature on the keywords in English (German, French):

- Euclidean algorithm* (Euklidischer Algorithmus, L'algorithme d'Euclide),
- Continued fraction (Kettenbruch*, Fraction continue*)
- Fibonacci number (Fibonacci-Folge, Suite de Fibonacci)

*) *Marked as an article worth reading*

More informations can be found at **Wolfram Mathworld** under the keywords:

- Continued Fraction, Euclidean Algorithm, Fibonacci Number

Extensive explanations on the topic of "rectangle dissections", especially in connection with the golden section, can be found in the books of Hans Walser, including

- Walser, Hans (2012): *Fibonacci. Zahlen und Figuren.* Edition am Gutenbergplatz, Leipzig (in German)
- Walser, Hans (2013): *Der Goldene Schnitt.* 6., bearbeitete und erweiterte Auflage. Edition am Gutenbergplatz, Leipzig (in German)

A detailed description of the algorithms (including a program for displaying the dissection of rectangles) can be found on the following website:

- https://www.maths.surrey.ac.uk/hosted-sites/R.Knott/Fibonacci/cfINTRO.html

**References to the Stamps from Liechtenstein (in German)**
- Börgens, Manfred: *Mathematik auf Briefmarken* # 86, https://homepages-fb.thm.de/boergens/marken/briefmarke086.htm
- Egelriede, Dieter: *Mathematik aus Liechtenstein*, Techno-Thema 69, S. 13, 2013
- Schierscher, Georg: *Geopythafibotonpolyhypotesaeder! Matheliebe*, catalogue of the exhibition, Verlag Liechtensteinisches Landesmuseum, 2012

# Circles and Circular Rings

# 4

*The horizon of many people is like a circle with radius zero.*
*And they call that their point of view.*

*(David Hilbert, German mathematician, 1862–1943)*

## 4.1 The Number $\pi$—The Circumference and Area of a Circle

The number $\pi$ is *defined* as $\pi = \frac{u}{d}$, that is, the ratio of the circumference $u$ of a circle to its diameter $d$ (= double radius $r$).

Archimedes (287–212 B.C.) gave an estimate for $\pi$: $3\frac{10}{71} < \pi < 3\frac{1}{7}$. For this purpose he determined the circumference of a regular inscribed 96-sided figure and a regular circumscribed 96-sided figure as well.

Starting with a regular 6-sided figure (hexagon), he gradually doubled the number of vertices.

© Springer-Verlag GmbH Germany, part of Springer Nature 2021
H. K. Strick, *Mathematics is Beautiful*, https://doi.org/10.1007/978-3-662-62689-4_4

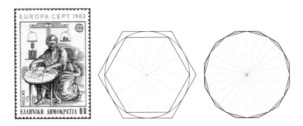

To calculate the area of a circle, he divided the circle into $n$ sectors and arranged the "pie slices" in such a way that (more and more precisely) the shape of a rectangle with width $\frac{u}{2}$ and height $r$ was formed.

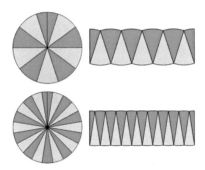

The area $A$ of the figure made of pie slices can then be approximately calculated from the width $\frac{1}{2} \cdot u$ and the height $r$. The formula for the area of a circle therefore results from the definition of the number $\pi$.

---

**Definition/Formula**

**Area $A$ of a circle**

The number $\pi$ is *defined* as the ratio of the circumference $u$ of the circle to its diameter $d$.

For the circumference $u$ of a circle we therefore have $u = \pi \cdot d = 2\pi \cdot r$.

For the area $A$ of a circle we have: $A = \frac{1}{2} \cdot u \cdot r = \frac{1}{2} \cdot d \cdot r$, that is, $A = \pi \cdot r^2 = 3.14159\ldots \cdot r^2$. ◄

The Chinese mathematician Zu Chongzhi (425–500) proceeded in a similar way to Archimedes. By continuously doubling the number of vertices up to the regular 24,576-sided figure, he was able to determine the number $\pi$ to seven decimal places. Stamps from China (1955) and from Hong Kong (2015) remind us of this achievement.

It was not until the fifteenth century that this accuracy was surpassed by the last great mathematician of the Islamic Middle Ages, Jamshid al Kashi (1380–1429, see the Iranian stamp from 1979), who was able to calculate the number $\pi$ to 16 decimal places by calculations on a regular 805,306,368-sided figure.

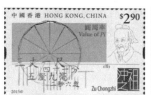

## 4.2　Circular Rings (Annuli)

According to the formula for calculating the area, a circle with radius $r = 1$ (unit) has the area $A_1 = 1\pi$ (square units). A circle with double radius, that is, $r = 2$, has the area $A_2 = 4\pi$, and so on.

The areas of the circular rings, that is, the area between two adjacent concentric circles, are obtained by difference:

$$A_{R12} = 4\pi - 1\pi = 3\pi, \quad A_{R23} = 9\pi - 4\pi = 5\pi, \quad A_{R34} = 16\pi - 9\pi = 7\pi \text{ and so on.}$$

---

### Rule
**Sequence of the circular rings of adjacent circles**

From the sequence of circles whose radii are natural numbers we come to the sequence of circular rings. For these, the following applies:

The sequence of the areas of the circular rings of adjacent circles corresponds to the sequence of *odd* multiples of the number $\pi$. ◄

---

### Suggestions for Reflection and for Investigations
**A 4.1**: Since the area of a circular ring of width 1 length unit is an odd multiple of $\pi$, an even multiple of $\pi$ results when the total area of two circular rings is considered.

Various combinations are possible. Continue the following statement.

There is 1 possible combination each for the areas $4\pi$ and $6\pi$, …

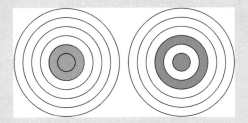

… 2 possible combinations each for the areas $8\pi$ and $10\pi$, …

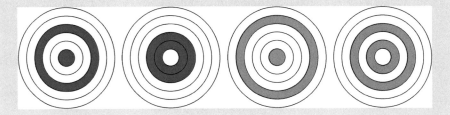

… 3 possible combinations for the area $12\pi$.

**A 4.2**: Since the area of a circular ring of width 1 is an odd multiple of $\pi$, an odd multiple of $\pi$ is also obtained if you calculate the total area of one, three, five, … circular rings, generally of an *odd* number of circular rings.

For example, the following figures show an area of $9\pi$ and of $11\pi$. How many possibilities are there to represent an odd multiple of $\pi$ using three circular rings?

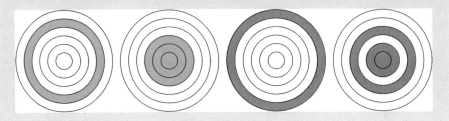

**A 4.3**: If you look at an even number of circular rings, you get an even multiple of π. How many possibilities are there to represent an even multiple of π using four circular rings?

**A 4.4**: The figures shown in the following illustrations consist of subareas each of the same size, for example, in the illustration on the left the green and the light blue colored areas are the same size, etc.

Why is it not possible to draw such figures with radii that are multiples of natural numbers?

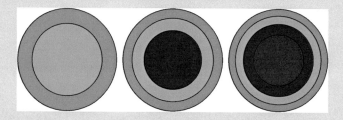

**A 4.5**: Explain: The circular ring figures in the following illustrations are *approximate* solutions for the problem "light blue" = "green" in **A 4.4.**

*Note:* In the figures shown, 5 of 7 and 12 of 17, and 29 of 41 circular rings are colored light blue, the others green - in the second and third figure the intermediate rings are not drawn.

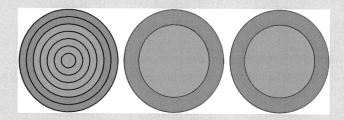

**A 4.6**: Find approximate solutions for the other two circle figures in **A 4.4** analogously to **A 4.5**.

**A 4.7**: Why is it difficult for us to realize that the differently colored areas in the following figures have the same size?

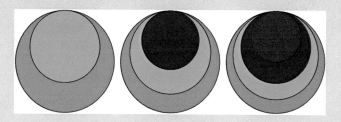

**A 4.8** Each of the figures in the following illustrations contain areas of equal size. Explain. Is it possible to assign them to specific area sizes ?

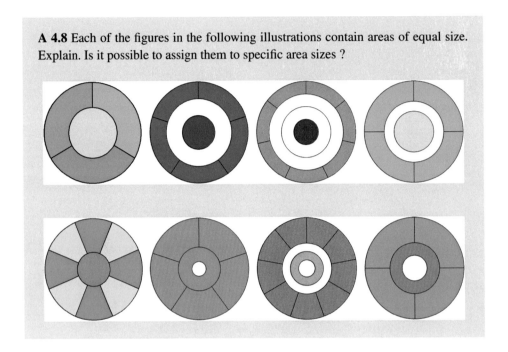

## 4.3    Shifted Semicircles

You can cut the circular rings along a diameter and move the two halves horizontally in opposite directions.

In the following nine figures seven concentric circles with the radii 1, 2, 3, 4, 5, 6, 7 are drawn and the semicircles are moved by 1, 2, 3, …, 8, 9.

If the cutting line (diameter) is disregarded, different areas are created (recognizable by the same coloring). The number of different areas is equal to the number of units by which the semicircles have been moved.

**Suggestions for Reflection and for Investigations**
**A 4.9**: All figures above have an area of $49\pi$. What is the ratio of the differently colored subareas?

## Ornaments of Circular Ring Figures

Following an idea of Hans Walser, several circular ring figures are drawn in a row, cut along a diameter and then moved horizontally.

**Example: Circular Ring Figures of Three Figures with Six Concentric Circles**
In case of shifting 1 unit, a continuous spiral pattern results (because of the contrast only one half is colored). If you move 2 units you get two infinitely long, intertwined meandering bands.

While with the shift of 3 units you get three infinite, intertwined bands. With the shift of 4 units the ornaments consists of infinitely many short elements.

In order to describe the ornament, which is shifted by 5 units, more precisely, one would have to increase the number of concentric circles lined up (then one will notice that it is a single, back-and-forth-going band). In case of a shift by 6 units, six infinitely long bands result.

**Suggestions for reflection and for investigations**

**A 4.10**: If one compares the graphics of the example with the results of Hans Walser for ten concentric circles, one can see that there are still some things that could be explored, for example:

a) Which characteristics do you find if you consider circular ring figures with *three* concentric circles, which are shifted by 1, 2, 3 units?

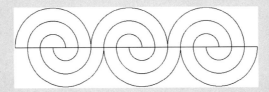

b) Which characteristics do you find if you consider circular ring figures with four concentric circles, which are shifted by 1, 2, 3, 4 units?

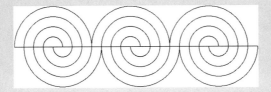

c) Which characteristics do you find if you consider circular ring figures with five concentric circles, which are shifted by 1, 2, 3, 4, 5 units?

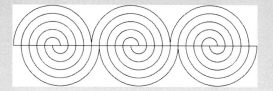

## 4.4   Braided Bands

Circular rings can also be matched together in other ways. For example, you can create a braiding pattern.

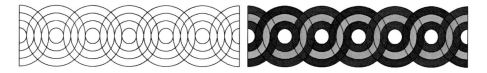

**Suggestions for Reflection and for Investigations**
**A 4.11**: Discover your own patterns with circular rings!

## 4.5   Tracks

If you cut a circular ring figure into two halves and insert parallel lines between the two parts of the figure, you get a geometric figure, which you could call a *track*.

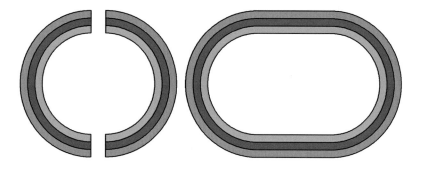

Tracks are also formed when one cuts out sectors of two circular ring figures with *different* radii where the angles complement to 360° and then connects the figures with straight sections.

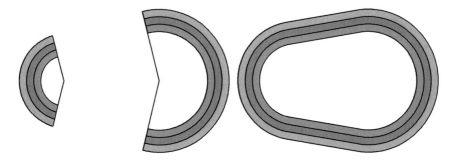

The necessary construction of (external) tangents to two circles can be seen in the following figures: The radii of the two circles are designated by $r$ and $R$ ($r \leq R$), the connecting line segment between the circle centers by $e$.

First you draw the two circles, then two auxiliary circles: one around the center $M_2$ of the larger circle, with radius $R - r$ (red) and one around the center of the line segment $M_1 M_2$, with radius $\frac{1}{2} \cdot e$ (green). According to Thales' theorem, a right angle exists at each of the intersections $S_1$ and $S_2$ of these two circles.

The rays outgoing from the center $M_2$ through the intersections $S_1$ and $S_2$ respectively cut the larger circle at the points of tangency. The points of tangency at the smaller circle are obtained by parallel translation of the line segments $M_1 S_1$ and $M_1 S_2$ respectively.

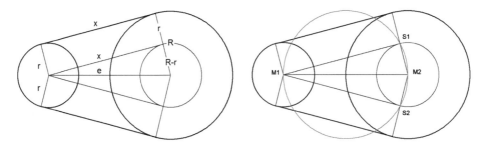

**Suggestions for Reflection and for Investigations**

**A 4.12**: Set up a formula to calculate the length of a track.

*Hint:* To calculate the length of a track, you need the length of the line segment $x$, which can be calculated by applying the Pythagorean theorem $x^2 + (R - r)^2 = e^2$, as well as the two arc length. For this purpose, one determines the two angles of the circle sectors at $M_1$ and $M_2$ using the relationship $\sin(\alpha) = (R - r)/e$ and from this, the acute angle $\alpha$ in the triangle $M_1 M_2 S_1$ and thus the two angles $180° - 2\alpha$ respectively $180° + 2\alpha$ of the two sectors.

**A 4.13**: As is well known (e.g., in a 400 m race) the runners on the outer track are given a so called stagger so that the running distances are equally long. Could such a compensation also be achieved by using a track in the form of a sideways figure eight?

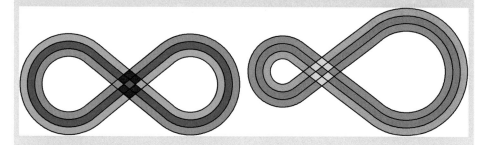

*Hint:* The construction of crossing tangents to two circles can be seen in the following figure.

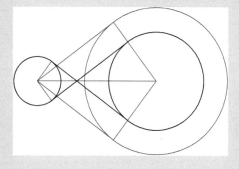

## 4.6   References to Further Literature

On **Wikipedia** you can find further information and literature on the keywords in English (German, French):

- Circular ring/Annulus (Kreisring, Couronne)
- Tangent lines to circles (Kreistangente, Cercle)

More informations can be found at **Wolfram Mathworld** under the keywords:

- Circle, Annulus, Concentric Circles, Circle Tangent Line, Circle-Circle Tangents

A representation of the ornaments from shifted semicircles can be found on the website

- https://www.walser-h-m.ch/hans/Miniaturen/A/Arch_Spiralen/Arch_Spiralen.pdf

**Literature reference to the shifted semi-circles**

- Walser, Hans: EAGLE-MALBUCH Formen und Farben: Geometrische Figuren zum Ausmalen, Edition am Gutenbergplatz Leipzig, 2015

# Pentominoes and Similar Puzzles

**5**

*The game is like a rest, and since you cannot work continuously, you need to rest.*

*(Aristotle, Greek philosopher, 384–322 B.C.)*

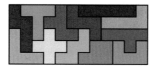

## 5.1    Simple Polyominoes

In reference to the word *domino* the American mathematician Solomon W. Golomb created in 1953 the terms *tromino, tetromino, pentomino, hexomino* … and the generic term *polyomino* for jigsaw pieces of this type. They were made popular by the monthly column of Martin Gardner in the *Scientific American*.

- **Dominoes:** A domino is created by putting two equal-sized squares together so that they have one square side in common – there is only one type here.

© Springer-Verlag GmbH Germany, part of Springer Nature 2021
H. K. Strick, *Mathematics is Beautiful*, https://doi.org/10.1007/978-3-662-62689-4_5

- **Trominoes:** If you put three equal-sized squares together, two different types of pieces are possible.

Domino and tromino gaming pieces can be laid out in different positions by turning them by 90°, 180°, and 270°. With these pieces no other variations are possible if you flip the forms.

In the following, we will not distinguish between all these variants; rather, we will imply the possibility that the polyomino pieces can be turned and flipped.

- **Tetrominoes:** Five types of gaming pieces are possible if you put four squares together. Usually the five types are designated by the letters I, O, L, T, and S.

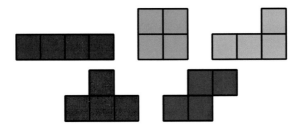

**Suggestions for Reflection and for Investigations**

**A 5.1**: With the five tetrominoes you can cover an area of 20 area units.

Explain: It is not possible to use the five tetrominoes to tessellate a rectangle with an area of 20 area units, neither a $2 \times 10$ rectangle nor a $4 \times 5$ rectangle.

*Hint:* Color the 20 squares of the two rectangles and the five tetrominoes alternately dark and light like a chessboard.

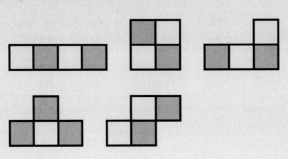

**A 5.2**: With a double set of tetrominoes you can cover a rectangular area of 40 units. Find more examples.

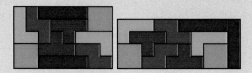

**A 5.3**: Explain: It is not possible to cover a square with $4 \times 4 = 16$ cells with four different tetrominoes.

**A 5.4**: Is it possible to cover a rectangular area with $4 \times 6 = 24$ cells with *at least* one element of each type?

**A 5.5**: The number 28 is the smallest triangular number divisible by 4 (see Sect. 2.2). Can a stair figure of 28 squares be tiled with the five tetrominoes?

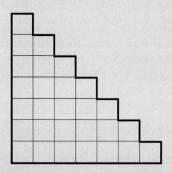

**A 5.6**: The numbers 16 and 36 are square numbers divisible by 4, which can be illustrated in the form of axially symmetric triangular figures. How can they be laid out with tetrominoes?

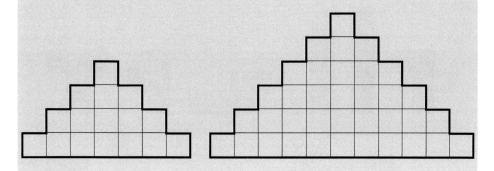

## 5.2    Pentominoes

There are twelve ways of putting five squares together to get a piece of the pentomino game. Usually these forms are designated with the letters F, I, L, N, P, T, U, V, W, X, Y, Z (see Fig. 5.1).

> **Suggestions for Reflection and for Investigations**
> **A 5.7**: Color the twelve pentominoes alternately dark and light like a chessboard (similar to the tetrominos in **A 5.1**). What differences do you see?

### 5.2.1    Tessellation of Rectangles with Pentominoes

With the twelve pieces of the pentomino game you can cover an area of 60 units.

There are six essentially different ways of drawing a rectangle with integral side lengths and an area of 60 units. As at least a height of 3 length units is required for laying out the pentominoes F, T, L, W, X, and Z, the $1 \times 60$ rectangle and the $2 \times 30$ rectangle cannot be considered for a tessellation with pentominoes.

Therefore: the $3 \times 20$ rectangle, the $4 \times 15$ rectangle, the $5 \times 12$ rectangle, and the $6 \times 10$ rectangle remain.

- For tessellation of a rectangular area with $3 \times 20 = 60$ fields only the following two essentially different possibilities exist.

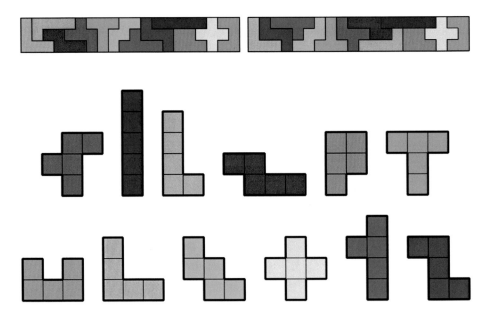

Fig. 5.1  The twelve possible pentominoes

In contrast the number of possible tessellations of the other rectangles is considerably larger:

- For the tessellation of $4 \times 15$ rectangles there are 368 possibilities, *one* of which is illustrated:

- For the tessellation of $5 \times 12$ rectangles there are 1,010 possibilities, *one* of which is illustrated:

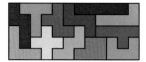

- For the tessellation of $6 \times 10$ rectangles there are 2,339 possibilities, *one* of which is illustrated:

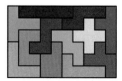

**Suggestions for Reflection and for Investigations**

**A 5.8**: Although there are many possibilities of tessellation, it is not easy to cover the 60 squares with exactly the twelve tiles. Apart from the examples shown, find at least one more possibility to tile a $4 \times 15$ rectangle, a $5 \times 12$ rectangle, or a $6 \times 10$ rectangle with the twelve pentominoes.

**A 5.9**: Determine all seven ways to lay out a rectangular area of $3 \times 5 = 15$ fields with three different pieces of the pentomino game.

Also for other rectangular forms, whose area is divisible by 5, there are numerous ways of laying out these with the respective number of pentominoes, whereby none of the pieces may be used twice (see the following table).

| Rectangle | 4 x 5 | 5 x 5 | 6 x 5 | 7 x 5 | 8 x 5 |
|---|---|---|---|---|---|
| Number of possibilities | 50 | 107 | 541 | 1396 | 3408 |
| Rectangle | 9 x 5 | 10 x 5 | 11 x 5 | 3 x 10 | 4 x 10 |
| Number of possibilities | 5902 | 6951 | 4103 | 145 | 2085 |

**Suggestions for Reflection and for Investigations**

**A 5.10**: In addition to the following examples, find at least one more way of tiling the rectangles with integral areas with different pentominoes.

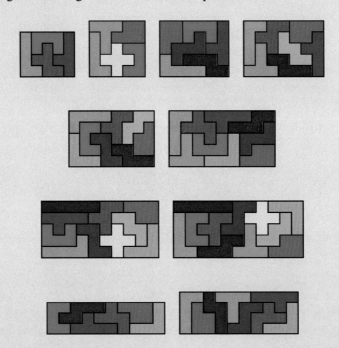

**A 5.11**: The following figures have an area of 60 units. Apart from the examples shown, find at least one more way of tessellating these figures with different pentominoes.

a) There are a total of 65 ways of tessellating a square area of 8 × 8 = 64 cells with the twelve different pentominoes, in which the four central cells may not be covered.

b) There are a total of 2,170 ways of tessellating a square area of 8 × 8 = 64 cells with the twelve different pentominoes, where the four cells in the corners cannot be covered.

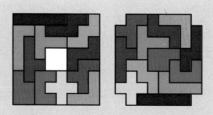

**A 5.12**: The following figures show two possible grids for a calendar sheet of a month. After placing six suitable pentominoes, one cell remains free – today's date! Choose a date and cover the remaining cells. Develop your own grid.

| 1 | 2 | 3 | 4 | 5 | 6 | 7 | 8 |
|---|---|---|---|---|---|---|---|
| 9 | 10 | 11 | 12 | 13 | 14 | 15 | 16 |
| 17 | 18 | 19 | 20 | 21 | 22 | 23 | 24 |
| 25 | 26 | 27 | 28 | 29 | 30 | 31 | |

| | 1 | 2 | 3 | 4 | 5 | |
|---|---|---|---|---|---|---|
| 6 | 7 | 8 | 9 | 10 | 11 | 12 |
| 13 | 14 | 15 | 16 | 17 | 18 | 19 |
| 20 | 21 | 22 | 23 | 24 | 25 | 26 |
| | 27 | 28 | 29 | 30 | 31 | |

## 5.2.2 Tessellation of Enlarged Pentomino Figures by Pentominoes

It is also an interesting challenge to tessellate the enlargements of the individual pentominoes and then cover them only with the *other* pentominoes. The enlarged figures are three times as high and three times as wide as the pieces themselves. For the figures with $3^2 \cdot 5 = 45$, cells nine of the eleven remaining pentominoes must be used. The number of possible tessellations can be taken from the following table.

| Pentominoes | F | I | L | N | P | T |
|---|---|---|---|---|---|---|
| Number of possibilities | 125 | 19 | 113 | 68 | 497 | 106 |
| Pentominoes | U | V | W | X | Y | Z |
| Number of possibilities | 48 | 63 | 91 | 15 | 86 | 131 |

**A 5.13**: In the following figures for each of the enlarged pentomino shapes one example of a possible tessellation is shown - covered by nine of the remaining eleven pentominoes. Please find at least one more possibility for tessellation of these figures.

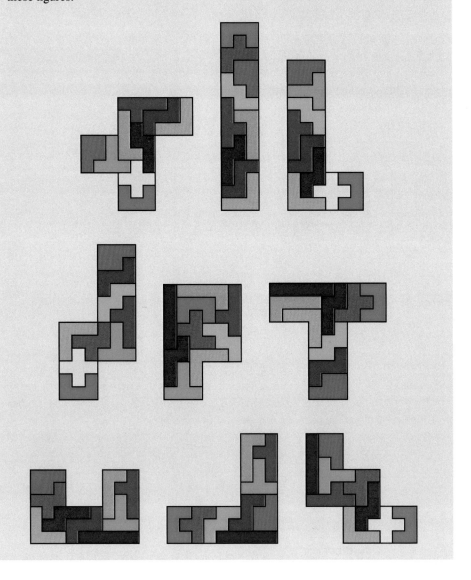

### 5.2.3 Tessellation of Triangular Figures Using Pentominoes

The elements of the sequence of the subtotals of the first $n$ natural numbers are called triangular numbers (see Chap. 2). Some of the triangular numbers are divisible by 5, which means that the corresponding triangular figures can be laid out with pentominoes.

We will only consider here the tessellation of triangular figures whose surface area does not exceed 60 area units and check whether and how often the area can be covered by different pentominoes.

There are four possible forms, namely with $1 + 2 + 3 + 4 = 10$ area units, with $1 + 2 + 3 + 4 + 5 = 15$ area units, with $1 + 2 + 3 + 4 + 5 + 6 + 7 + 8 + 9 = 45$ area units, and with $1 + 2 + 3 + 4 + 5 + 6 + 7 + 8 + 9 + 10 = 55$ area units (see the following figures).

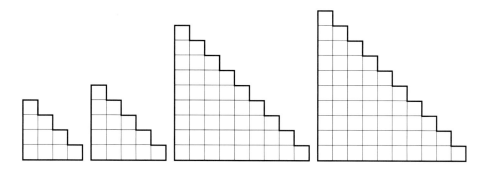

**Suggestions for Reflection and for Investigations**

**A 5.14**: Color the squares of the four triangular figures shown, like a chessboard, alternately dark and light. What special features do you discover?

**A 5.15**: Determine all possibilities to cover the triangular figure of 10 area units with two different pentominoes or the triangular figure of 15 area units with three different pentominos. One example is shown in the following figure.

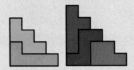

**A 5.16**: Is it possible to lay out the triangular figures of 45 area units or 55 area units with different pentominoes?

Another triangular figure is obtained by stacking rectangles with odd-numbered sides as shown in the following figure. As explained in Chap. 2, the total number of squares contained in the figure is a square number, since the sum of the first consecutive odd natural numbers is always a square number.

The only square number divisible by 5 up to 60 is the number 25, represented as the sum of the first five odd natural numbers: $1 + 3 + 5 + 7 + 9 = 5^2$ (see the following figure).

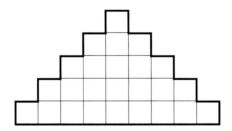

**Suggestions for Reflection and for Investigations**

**A 5.17**: Determine all ways of covering the triangular figure considered above (with an area of 25) with five different pentominoes. Note that some possibilities result directly from the solutions in A 5.14 (see the following figures).

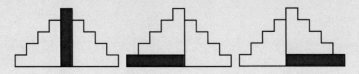

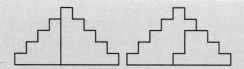

**A 5.18**: The sum of consecutive square numbers can also be represented by a rotated square figure. The smallest sum of consecutive square numbers divisible by 5 is

$$3^2 + 4^2 = 9 + 16 = (1 + 3 + 5) + (1 + 3 + 5 + 7) = 5^2.$$

A tessellation of this shape is shown in the picture on the right. Does another possibility of tessellation exist?

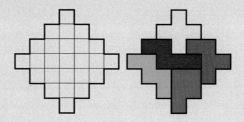

## 5.3    Hexominoes

There are 35 ways to put six squares together to form a hexomino piece. The following figure shows the eleven hexominoes that could be folded up to form a cube; the remaining 24 forms may be found by yourself.

In principle, similar tasks can be developed for hexominoes as for pentominoes. However, since the number of hexominoes is much larger, the number of solutions will certainly be very large.

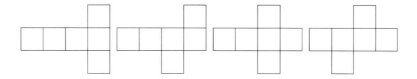

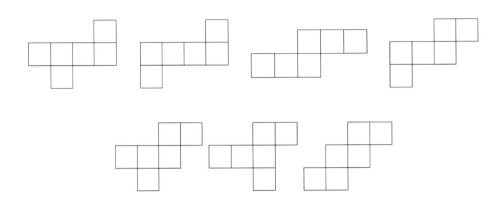

**Suggestions for Reflection and for Investigations**

**A 5.19**: Determine the 24 hexominoes that are *not* suitable for nets of a cube.

**A 5.20**: Which of the hexominoes are made of which pentominoes by adding another square? Which of the pentominoes are made of which tetrominoes by adding another square?

## 5.4     References to Further Literature

On **Wikipedia** you can find further information and literature on the keywords in English (German, French):

- Pentomino (Pentomino, Pentomino)
- Polyomino (Polyomino, Polyomino)

Extensive informations can be found at **Wolfram Mathworld** under the keywords:

- Pentomino, Polyomino

Recommended literature:

- Golomb, Solomon W. (1994, 2$^{nd}$ edition), *Polyominoes – Puzzles, Patterns, Problems, and packings*, Princeton University Press, Princeton

A detailed description of the possibilities is contained in the websites.

- https://isomerdesign.com/Pentomino/
- https://gp.home.xs4all.nl/PolyominoSolver/Polyomino.html

# Curve Stitching

# 6

*Whoever works on the wrong thread destroys the whole textile.*

*(Confucius, Chinese philosopher, 551–479 B.C.)*

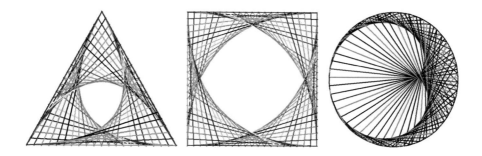

You can easily create apparently curved lines by stretching threads. To do this, you draw the basic figure (circle, equilateral triangle, square, regular polygon) on a board, hammer in nails at regular intervals, which are then joined together according to certain rules. In this way circles, parabolas and other curves are created in our brain, although we actually only see tangents to these curves.

## 6.1    Circle as Basic Figure—Sides and Diagonals in Regular Polygons

If nails are driven in on a circular line at $n$ regular intervals and each nail is connected with all the other nails, a regular $n$-sided figure is formed with $n$ sides and $\frac{1}{2} \cdot n \cdot (n-3)$ diagonals (seSect. 1.3).

© Springer-Verlag GmbH Germany, part of Springer Nature 2021
H. K. Strick, *Mathematics is Beautiful*, https://doi.org/10.1007/978-3-662-62689-4_6

The following figures show various regular *n*-sided figures, which are obtained by doubling the number of vertices from a regular 6-sided or 8-sided figure: A regular 12-sided figure with 12 sides and 54 diagonals, a regular 16-sided figure with 16 sides and 104 diagonals, and a regular 24-sided figure with 24 sides and 252 diagonals:

As the number of vertices increases, the impression arises in our mind that circles have been drawn inside the figures. These circles do not really exist. What we actually see are tangents to the circles.

To check this, consider the diagonal of one vertex to the opposite vertex. This diagonal is the axis of symmetry for every two diagonals starting from the same point. The next two adjacent diagonals are tangents to the innermost circle, the next two are tangential to the second innermost circle, and so on.

From the graphic you can see that the radius of the innermost (imaginary) circle is *approximately* half the size of the side length of the regular polygon.

The radius of the second (imaginary) circle is not quite twice as large as the radius of the innermost circle, the radius of the third circle is not quite three times as large; the increase in radius from inside to outside is somewhat smaller in each case.

In Chap. 1 we explained which vertices of a regular *n*-sided figure have to be connected with which vertices in order to create a *n*-pointed star, so that only *one* thread is needed. Even if here only some of the diagonals are needed, circles apparently develop inside the figure, see the following illustrations of the stars {19/8}, {20/7}, {21/8}.

## 6.2     Square as Basic Figure

### 6.2.1    Special Star Figures in a Square

If you choose the vertices of a square and other points in between, you get special star figures.

For example, you get an 8-pointed star {8/3} by connecting the vertices and midpoints of the sides, if you choose the sequence of points A–D–G–B–E–H–C–F–A, that is, connecting each of the eight points with the third next one. If one colors the subareas of the figure, again a especially beautiful star is created.

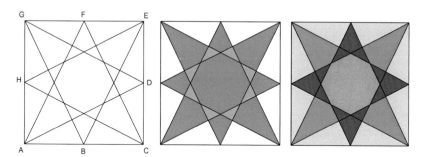

The special feature of this star figure: If you choose 60 units as side length of the square and place the origin of the coordinate system in a "corner" or in the center of the figure, then all appearing intersections have integral coordinates and all subareas are integers.

**Suggestions for Reflection and for Investigations**
**A 6.1**: In the following figures, various grid lines are drawn in the 8-pointed figure, from which the integral coordinates of the intersection points can then be read off. The side length of the surrounding square is 60 units.

Define the coordinates of the intersection points of the figure and calculate the subareas.

What are the ratios of the side lengths which can be found in the right-angled triangles?

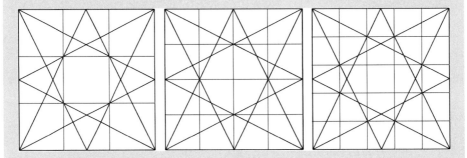

**A 6.2**: Determine in the same way as **A 6.1** the coordinates of the intersections of the following 12-pointed star.

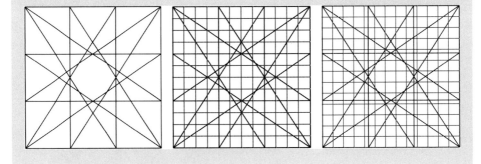

## 6.2.2  Parabolas in a Square

If you divide the left vertical and the lower horizontal side of a square into ten equally sized sections, number the subdivision points 0, 1, 2, …, 10 and connect.

- point No. 10 of the vertical side with point No. 1 of the horizontal side,
- point No. 9 of the vertical side with point No. 2 of the horizontal side,
- point No. 8 of the vertical side with point No. 3 of the horizontal side,
- …
- point No. 1 of the vertical side with point No. 10 of the horizontal side,

then the thread pattern in Fig. 6.1a results.

It looks as if a curved line has been created here – it is a parabola. But actually only tangents of the parabola have been drawn.

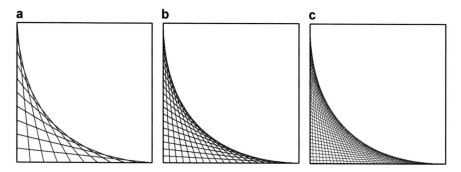

**Fig. 6.1   a–c** Apparent parabolas in a square

If the number of sections on the two square sides is increased (e.g., 20 sections or 40 sections, see Fig. 6.1b, c), the impression that a curve has actually been created is reinforced.

That the curve is not a quarter circle, can be seen in the following figure.

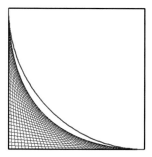

The following thread patterns are created when you create four of these parabolic "figures" in a square or when colored threads are used.

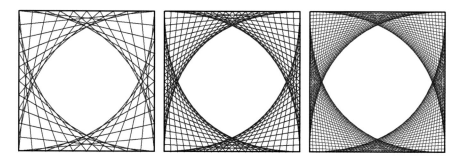

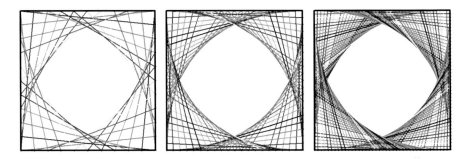

Instead of using a square as a frame, you can also use two perpendicular (coordinate) axes (or four squares put together to form a larger square).

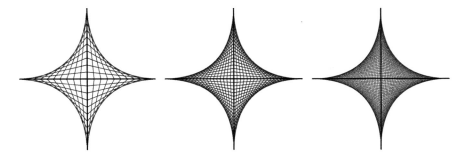

As a basic figure you can also use regular polygons or axes with special angles. The threads are stretched between two adjacent sides or axes:

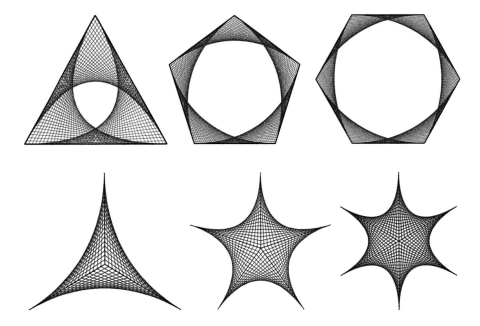

## 6.3     Digression: Envelope of a Family of Curves

In higher mathematics, these curves generated in our brain are called **envelopes** (or envelope curves). Since thread patterns are being looked at here, these are envelopes of a family of straight lines.

In order to determine the equation of such an envelope curve, one must analyze the functional equation of the family of straight lines. This goes beyond the methods usually dealt with in school lessons.

### 6.3.1   Examples of Families of Straight Lines

In differential calculus one considers tangents to the graphs of functions.

The slope of a tangent line at a point $(a, f(a))$ of a graph is calculated using the first derivative at this point. The equation of such a tangent line $t_a$ is given by:

$$t_a(x) = f'(a) \cdot (x - a) + f(a)$$

**Example 1: Quadratic Function**

For a quadratic function $f$ with $f(x) = x^2$ ... we get:

$$t_a(x) = 2a \cdot (x - a) + a^2 = 2ax - a^2$$

In Fig. 6.2a the family of straight lines of the tangents for $-4 \leq a \leq +4$ and increment $\Delta a = 0.2$ is drawn on the graph of the parabola. Although the parabola itself was not drawn, nevertheless one gets the impression that this graph is present.

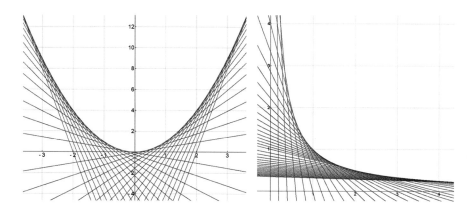

**Fig. 6.2** Tangents to a parabola and a rectangular hyperbola

**Example 2: Reciprocal Function**

For the reciprocal function $f$ with $f(x) = \frac{1}{x}$, whose graph is also called a rectangular hyperbola, the equation of the tangent line is given by:

$$t_a(x) = -\frac{1}{a^2} \cdot (x - a) + \frac{1}{a} = -\frac{1}{a^2}x + \frac{2}{a}$$

In Fig. 6.2b the tangents for $0.1 \le a \le 2$ with increment $\Delta a = 0.1$ and for $2 \le a \le 6$ with increment $\Delta a = 0.2$ are drawn.

### 6.3.2  Determining the Equation of the Enveloping Parabola

In the following, we describe a method which does not exceed the scope of knowledge of mathematics lessons at the high school level.

The figures in Fig. 6.1a–c obviously have an axis of symmetry which runs through the origin and a vertex of the curve yet to be determined. The vertex is located approximately where the two "central" straight lines of the family intersect.

---

**Calculation for 10 Divisions**

You look at the two "central" straight lines

$g_6$ which runs through $(0, 6)$ and $(5, 0)$: $y = \frac{0-6}{5-0} \cdot (x - 0) + 6 = \frac{-6}{5} \cdot x + 6$.

$g_5$ which runs through $(0, 5)$ and $(6, 0)$: $y = \frac{0-5}{6-0} \cdot (x - 0) + 5 = \frac{-5}{6} \cdot x + 5$.

For the $x$-coordinate of the point of intersection applies:

$$\frac{-6}{5} \cdot x + 6 = \frac{-5}{6} \cdot x + 5 \Leftrightarrow \left(\frac{6}{5} - \frac{5}{6}\right) \cdot x = 1 \Leftrightarrow x = \frac{30}{11} \approx 2.73$$

For the $y$-coordinate of the point of intersection applies: $y = \frac{-6}{5} \cdot \frac{30}{11} + 6 = \frac{30}{11}$.

---

**Calculation for 20 Divisions**

You look at the two "central" straight lines.

$g_{11}$ which runs through $(0, 5.5)$ and $(5, 0)$: $y = \frac{0-5.5}{5-0} \cdot (x - 0) + 5.5 = \frac{-5.5}{5} \cdot x + 5.5$

$g_{10}$ which runs through $(0, 5)$ and $(5.5, 0)$: $y = \frac{0-5}{5.5-0} \cdot (x - 0) + 5.5 = \frac{-5}{5.5} \cdot x + 5$.

For the $x$-coordinate of the point of intersection applies:

$$\frac{-5.5}{5} \cdot x + 5.5 = \frac{-5}{5.5} \cdot x + 5 \Leftrightarrow \left(\frac{5.5}{5} - \frac{5}{5.5}\right) \cdot x = 0.5 \Leftrightarrow x = \frac{13.75}{5.25} \approx 2.62$$

For   the   $y$-coordinate   of   the   point   of   intersection   applies:
$y = \frac{-5.5}{5} \cdot \frac{13.75}{5.25} + 5.5 \approx 2.62$.

**Generalization of the Calculation**

You look at the two "central" straight lines.

$g_r$ which runs through $(0,\ 5 + \Delta)$ and $(5, 0)$: $y = \frac{-5-\Delta}{5} \cdot x + 5 + \Delta$.

$g_{r-1}$ which runs through $(0, 5)$ and $(5 + \Delta,\ 0)$: $y = \frac{-5}{5+\Delta} \cdot x + 5$.

For the $x$-coordinate of the point of intersection applies:

$$\frac{-5 - \Delta}{5} \cdot x + 5 + \Delta = \frac{-5}{5+\Delta} \cdot x + 5 \Leftrightarrow \left(\frac{5+\Delta}{5} - \frac{5}{5+\Delta}\right) \cdot x = \Delta$$

$$\Leftrightarrow x = \Delta \cdot \frac{5 \cdot (5+\Delta)}{(5+\Delta)^2 - 5^2} = \Delta \cdot \frac{5 \cdot (5+\Delta)}{(5+\Delta-5) \cdot (5+\Delta+5)} = \frac{5 \cdot (5+\Delta)}{10 + \Delta}$$

As $\Delta \to 0$ x and $y$ converge to 2.5.

If the figure in Fig. 6.3a is now rotated 45° to the left, the resulting parabola lies symmetrically about the $y$-axis of a coordinate system (see Fig. 6.3b).

The points $(0, 10)$, $(2.5, 2.5)$, and $(10, 0)$ of the figure on the left are mapped to the following points by rotation: $P_1(-5\sqrt{2},\ 5\sqrt{2})$, $P_2(5\sqrt{2},\ 5\sqrt{2})$, $S(0,\ 2.5\sqrt{2})$.

The equation of the parabola with $y = ax^2 + b$ is then obtained from these points by comparing the coefficients:

$$y = \frac{\sqrt{2}}{20}x^2 + 2.5\sqrt{2}$$

To determine the equation of the original parabola from this, the graph must be rotated by 45° again (this time to the right).

After applying the formula (6.1), see below, a rotation of 45° to the right results, by applying $\sin(-45°) = -\frac{\sqrt{2}}{2}$ and $\cos(-45°) = \frac{\sqrt{2}}{2}$:

$$x' = \frac{\sqrt{2}}{2}(x+y) \text{ and } y' = \frac{\sqrt{2}}{2}(-x+y),$$

and solved for $x$, $y$:

$$x = \frac{\sqrt{2}}{2}(x' - y') \text{ and } y = \frac{\sqrt{2}}{2}(x' + y')$$

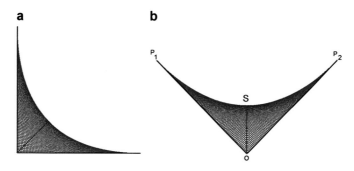

**Fig. 6.3  a, b** Turning the enveloping parabola by 45°

If we apply this to the initial equation $y = \frac{\sqrt{2}}{20}x^2 + 2.5\sqrt{2}$,

we will get $\frac{\sqrt{2}}{2} \cdot (x' + y') = \frac{\sqrt{2}}{20} \cdot \frac{1}{2} \cdot (x' - y')^2 + 2.5 \cdot \sqrt{2}$ and thus

$$20 \cdot (x' + y') = (x' - y')^2 + 100.$$

This equation for the envelope in Fig. 6.3a can explicitly be noted as:

$$x'^2 - 2x'y' + y'^2 - 20x' - 20y' + 100 = 0$$

---

**Formula**
### Rotation of a point around the origin
In a coordinate system, when a point $P\,(a, b)$ is rotated around the origin by a rotation with the angle $\alpha$ it moves to the point $P'\,(a', b')$. For the coordinates of $P'$ you have:

$$a = a \cdot \cos(\alpha) - b \cdot \sin(\alpha) \text{ and } b = a \cdot \sin(\alpha) + b \cdot \cos(\alpha) \qquad (6.1)$$

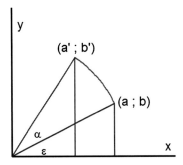

◄

### Derivation of the Formula (6.1)
For $r = |OP|$ according to Pythagoras's theorem: $a^2 + b^2 = |OP|^2$.

In the right-angled triangle defined by the coordinates of the point $P$, we have

$$a = r \cdot \cos(\varepsilon) \text{ and } b = r \cdot \sin(\varepsilon)$$

In the right-angled triangle defined by the coordinates of the point $P'$, we have:

$$a' = r \cdot \cos(\varepsilon + \alpha) \text{ and } b' = r \cdot \sin(\varepsilon + \alpha)$$

After applying the angle sum formulae for sin and cos

$$\sin(\varepsilon+\alpha) = \sin(\varepsilon) \cdot \cos(\alpha) + \cos(\varepsilon) \cdot \sin(\alpha) \text{ and}$$
$$\cos(\varepsilon+\alpha) = \cos(\varepsilon) \cdot \cos(\alpha) - \sin(\varepsilon) \cdot \sin(\alpha)$$

we get

$$a' = r \cdot \cos(\varepsilon + \alpha) = r \cdot \cos(\varepsilon) \cdot \cos(\alpha) - r \cdot \sin(\varepsilon) \cdot \sin(\alpha) = a \cdot \cos(\alpha) - b \cdot \sin(\alpha)$$

and

$$b' = r \cdot \sin(\varepsilon + \alpha) = r \cdot \sin(\varepsilon) \cdot \cos(\alpha) + r \cdot \cos(\varepsilon) \cdot \sin(\alpha) = b \cdot \cos(\alpha) + a \cdot \sin(\alpha).$$

## 6.4    Curves of Pursuit

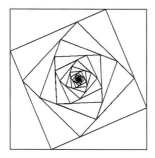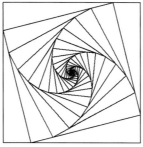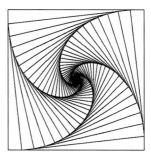

The illustrated figures contain a sequence of an infinite number of nested squares, which basic sides are divided into sections with the ratio 3:7 or 2:8 or 1:9.

These have been formed as follows: All four sides of the square are divided with the ratio indicated. All points of partition are then connected with the corresponding points on the adjacent sides – the resulting figure is again a square. In this square, the sides are also divided in the same ratio, and so on.

The name **curve of pursuit** can best be explained by the following story:

Four dogs are sitting at the vertices of a square when each of them notices an adjacent dog (clockwise or counter-clockwise direction). They run toward the neighboring dog, but after they have covered a part of the distance and all at the same time they notice that they have to change their direction, because the dog they are aiming at has changed its position in the meantime and so on.

While one can hardly speak of a "curve" in the above-mentioned graphs, this impression is intensified the closer the points of division lie to the vertices; in Fig. 6.4a, b the ratio of division is 1:19. Thread patterns in which the squares are shown in different colors are certainly impressive.

The apparently emerging curves are so-called **logarithmic spirals.** These spirals orbit the center of the square an infinite number of times; nevertheless, their length is finite – it is in fact as long as the side length of the square.

One of the special properties of logarithmic spirals is:

If you draw any straight lines through the center of the square, they will always intersect the spirals at the same angle, namely $45°$.

In practice, however, the thread patterns with pursuit curves quickly reveal problems with regard to the accuracy of the positions into where the nails must be driven in, as these are increasingly close together. When displayed on the computer screen, these

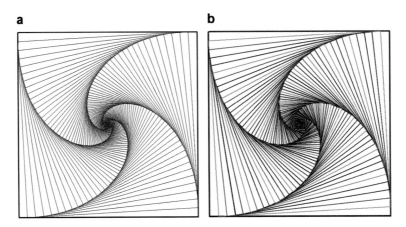

**Fig. 6.4   a, b** Logarithmic spirals as envelopes of the nested squares

problems are only encountered near the center of the squares where the spiral arms are turned in.

Figures with curves of pursuit can be designed in different ways, for example, by coloring the fields alternately or by drawing only the tracks that form the spirals, or by coloring the subareas created by the spirals.

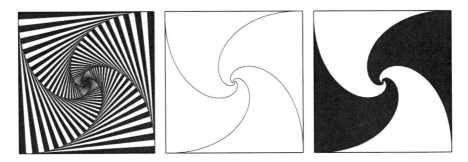

Logarithmic spirals are also created for other underlying polygons. It can be shown (by geometric series), that the length $\lambda_n$ of the spirals for a regular $n$-sided figure with side length $s$ is calculated as follows:

$$\lambda_n = \frac{s}{1 - \cos\left(\frac{360°}{n}\right)}$$

In the case of $n = 4$ results $\lambda_4 = s$, that is the spirals are – as indicated above – just as long as the sides of the square.

In an equilateral triangle the following applies to the length of the pursuit curve: $\lambda_3 = \frac{2}{3} \cdot s$, in a regular 5-sided figure applies $\lambda_5 \approx 1,447 \cdot s$, and in regular 6-sided figures the logarithmic spirals are twice as long as the sides: $\lambda_6 = 2 \cdot s$.

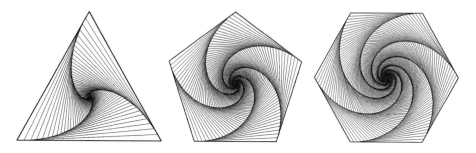

## 6.5    Circle as Basic Figure: Epicycloid

A **cycloid** is a curve traced by a point on a circle (a wheel) as it rolls along a straight line. If the moving circle rolls on another fixed circle, the resulting curve is called an **epicycloid.**

The type of curve created depends on the ratio of the radii of the fixed and the moving circles.

In the following we will look at the following examples:

1. The rolling circle and the fixed circle have the radius $r$.
2. The fixed circle has a radius of $2r$, the rolling circle the radius $r$.
3. The fixed circle has a radius of $3r$, the rolling circle the radius $r$.

In Fig. 6.5a–c, the center of the fixed circle is at the origin and is colored yellow, and the rolling circle is shown in blue, each in several different positions.

The curves can be described by the following parameter representations (see Fig. 6.6a–c):

$$x(t) = r \cdot [2\cos(t) - \cos(2t)]; \ y(t) = r \cdot [2\sin(t) - \sin(2t)]$$
$$x(t) = r \cdot [3\cos(t) - \cos(3t)]; \ y(t) = r \cdot [3\sin(t) - \sin(3t)]$$
$$x(t) = r \cdot [4\cos(t) - \cos(4t)]; \ y(t) = r \cdot [4\sin(t) - \sin(4t)]$$

Epicycloids can also be created by thread patterns:

If you connect the vertices of a regular $n$-sided figure (i.e., points on a circle) according to a certain rule, then such epicycloids are created in our brain.

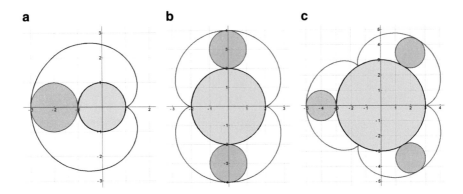

**Figure 6.5  a–c** Epicycloids with the fixed circle and the rolling circle in several different positions

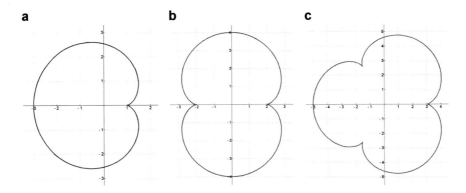

**Fig. 6.6  a–c** Graphs of the parameter representations

First we look at a regular 72-sided figure. The vertices are numbered from 1 to 72; the numbering is then continued so that vertex $k$ and vertex $72 + k$, $144 + k$, etc. are the same.

If you then connect vertex No. $k$ with

(1) Vertex No. $2k$ or (2) vertex No. $3k$ or (3) vertex No. $4k$

then one obtains the epicycloids described above as envelopes; the contours become sharper if the number of vertices is doubled to 144 (see, following figures).

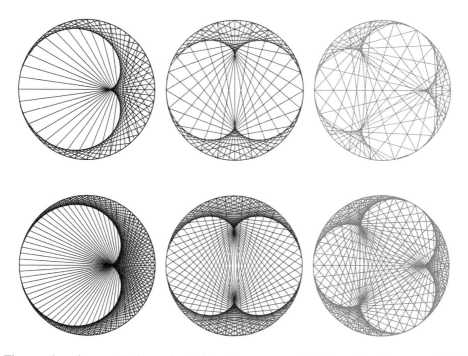

The envelope in case (1) is a epicycloid with one cusp; it is also referred to as **cardioid** (Greek *kardia, Engl. heart*), in case (2) it is a epicycloid with two cusps, which is called **nephroid** (Greek *nephros,* Engl. *kidney*).

## 6.6   Perpendicular Axes as Basic Figure: Astroids

The enveloping curve in the following two graphs looks like a parabola, but it is actually a so-called **astroid.**

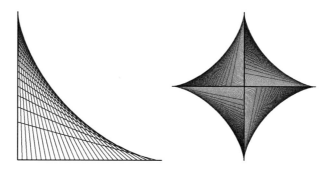

The first thread pattern is determined as follows:

On the $x$-axis, you mark points at equal distances. Then you take threads with a fixed length, attach them to the points of the $x$-axis and then determine corresponding points on the $y$-axis for the other end of the thread.

One can also imagine that the enveloping curve was created as follows:

A ladder with a certain length slides down a wall (the $y$-axis). For certain positions on the $x$-axis (which are at equal distances from each other) the corresponding point on the $y$-axis at which the ladder is leaning are marked.

The astroids are examples of so-called **hypocycloids.** These are created when a moving circle rolls *inside* a fixed circle.

In Fig. 6.7a the fixed circle has the radius $r$ and the rolling circle has the radius $\frac{1}{4}r$.

The parameter representation of the curve shown in Fig. 6.7b is:

$$x(t) = r \cdot \cos^3(t); y(t) = r \cdot \sin^3(t)$$

The curve can also be described by the equation

$$x^{\frac{2}{3}} + y^{\frac{2}{3}} = r^{\frac{2}{3}}.$$

The length of the astroid is with $6r$ slightly shorter than the circumference of the circle $(2\pi r)$.

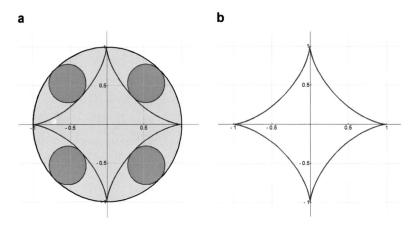

**Fig. 6.7**   a, b Example of an astroid

## 6.7    References to Further Literature

On **Wikipedia** you can find further information and literature on the keywords in English (German, French):

- Envelope (Einhüllende, Enveloppe)
- Pursuit curve (Radiodrome, Courbe du chien)
- Epicycloid (Epizykloide, Épicycloïde)

Under the keyword Curve stitching (Fadenbilder, Courbe à broder) you can find numerous suggestions for activities on the Internet, including the following book:

- Hale, Helen, Curve Stitching: Art of Sewing Beautiful Mathematical Patterns, Tarquin Pub, 2008

Extensive information can be found at **Wolfram Mathworld** under the keywords:

- Envelope, Pursuit Curve, Epicycloid, Cardioid, Astroid

# Calculating with Square Numbers—Number Cycles

# 7

*Every boy must learn arithmetic so that he can calculate his life, because all reason, especially the management of human things, is called arithmetic.*

*(Johann Gottfried von Herder, German poet and philosopher, 1744–1803)*

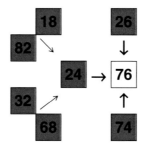

This chapter deals with a remarkable property of the final digits of square numbers which could be described by the term "square number cycles".

Before these number cycles are discovered in the following, we will show some ideas for mental calculation of square numbers. The research on square number cycles can also be done with the help of pocket calculators or a spreadsheet, so if you are not interested in the hints for mental arithmetic you can skip the first part of the chapter.

© Springer-Verlag GmbH Germany, part of Springer Nature 2021
H. K. Strick, *Mathematics is Beautiful*, https://doi.org/10.1007/978-3-662-62689-4_7

## 7.1 Calculating with Square Numbers

### 7.1.1 Calculating with Square Numbers: From One Square Number to the Next

As explained in Chap. 2, square numbers can be visualized by means of dot patterns.

To go from one square number to the next, you have to add a L-shaped form with an odd number of dots to the square of dots, that is, the difference between two consecutive square numbers $n^2$ and $(n + 1)^2$ is the odd number $2n + 1$.

The term $2n + 1$ can also be described as $n + (n + 1)$ as can be seen in the right figure below.

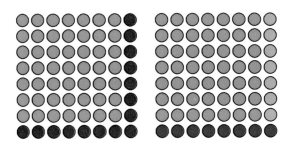

From this follows the first rule for calculating square numbers:

---

**Rule 1**

If you know the square $n^2$ of a natural number $n$, then you get the next square number $(n + 1)^2$ by adding the number $n$ and its successor $(n + 1)$ to the square number $n^2$:

$$(n + 1)^2 = n^2 + n + (n + 1) \blacktriangleleft$$

| | | | |
|---|---|---|---|
| $40^2 = 1600$ | 40 | 41 | $41^2 = 1600 + (40 + 41) = 1600 + 81 = 1681$ |
| $41^2 = 1681$ | 41 | 42 | $42^2 = 1681 + (41 + 42) = 1681 + 83 = 1764$ |
| $42^2 = 1764$ | 42 | 43 | $43^2 = 1764 + (42 + 43) = 1764 + 85 = 1849$ |
| ... | | | |

In principle, one can also calculate the further square numbers step by step from *one* known square number. Since then the sum of two, four, six, ... numbers has to be calculated, it is easier to multiply the mean value of the summands by the number of summands instead of adding up the consecutive numbers.

| $40^2 = 1600$ | 40 | 41 | $41^2$ = 1600 + (40 + 41)<br>= 1600 + 2 · 40,5 = 1600 + 81 = 1681 |
| | | 42 | $42^2$ = 1600 + (40 + 41) + (41 + 42)<br>= 1600 + 4 · 41 = 1600 + 164 = 1764 |
| | | 43 | $43^2$ = 1600 + (40 + 41) + (41 + 42) + (42 + 43)<br>= 1600 + 6 · 41,5 = 1600 + 249 = 1849 |
| ... | | | |

This method can also be applied in reverse steps.

| $50^2 = 1600$ | 50 | 49 | $49^2$ = 2500 − (50 + 49)<br>= 2500 − 2 · 49.5 = 2500 − 99 = 2401 |
| | | 48 | $48^2$ = 2500 − (50 + 49) − (49 + 48)<br>= 2500 − 4 · 49 = 2500 − 196 = 2304 |
| | | 47 | $47^2$ = 2500 − (50 + 49) − (49 + 48) − (48 + 47)<br>= 2500 − 6 · 48.5 = 2500 − 291 = 2209 |
| ... | | | |

## 7.1.2   Calculating with Square Numbers: A Special Rule for Square Numbers with the Final Digit 5

If you look at the squares of natural numbers with final digit 5, you will realize a special property:

| $n$ | $n^2$ | how to calculate | $n$ | $n^2$ | how to calculate |
|---|---|---|---|---|---|
| 5 | 25 | | 55 | 3025 | $50 · 60 + 5^2$ |
| 15 | 225 | $10 · 20 + 5^2$ | 65 | 4225 | $60 · 70 + 5^2$ |
| 25 | 625 | $20 · 30 + 5^2$ | 75 | 5625 | $70 · 80 + 5^2$ |
| 35 | 1225 | $30 · 40 + 5^2$ | 85 | 7225 | $80 · 90 + 5^2$ |
| 45 | 2025 | $40 · 50 + 5^2$ | 95 | 9025 | $90 · 100 + 5^2$ |

---

**Rule 2**

You get the square of a natural number with final digit 5 by adding the number 25 to the product of the next lower and the next higher multiples of ten:

$$(10 · k + 5)^2 = [10 · k] · [10 · (k + 1)] + 5^2 = 100 · k · (k + 1) + 5^2 \blacktriangleleft$$

This can easily be illustrated:

**Example: Justification of the rule for the calculation of $35^2$**

According to binomial formula we have:

$$35^2 = (30 + 5)^2 = 30^2 + 2 \cdot 30 \cdot 5 + 5^2$$

That means, the square with side length 35 units is composed of the square with side length 30, two $30 \times 5$ rectangles and a square with side length 5 units.

By moving one of the two rectangles, the square with side length 30 units and the two rectangles form a "large" rectangle with the side lengths 30 units and 40 units.

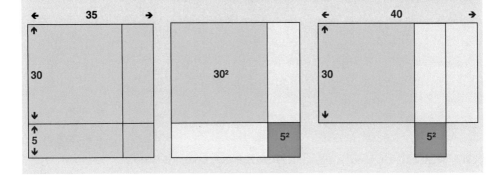

*Historical note:* This way of illustrating this binomial formula can be found in the *Elements* of Euclid (Book II, Proposition 4).

### 7.1.3   Calculating with Square Numbers: Using Equidistant Numbers

The geometric idea of factoring a term can also be used to calculate square numbers as follows:

A square with side length $n$ units can be divided into a square with side length $(n - 1)$ units, two $(n - 1) \times 1$ rectangles and a square with side length 1 unit.

By moving one of the rectangles you get an augmented $(n - 1) \times (n + 1)$ rectangle, so we get (see also the following figure):

$$n^2 = (n - 1) \cdot (n + 1) + 1^2$$

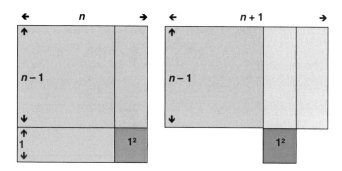

Analogously similar relationships apply for natural numbers equidistant to a number $n$ (see also the following figure):

$$n^2 = (n - 2) \cdot (n + 2) + 2^2$$

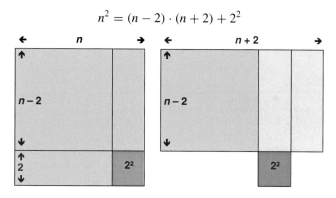

---

**Rule 3**

The square of a natural number $n$ is equal to the product of two numbers lying equidistantly (symmetrically) to $n$, which are $n - k$ and $n + k$ ($0 \leq k \leq n$), augmented by $k^2$:

$$n^2 = (n - k) \cdot (n + k) + k^2 \blacktriangleleft$$

In order to make especially elegant use of these relationships, one has to calculate the distance to the next *number of tens* – the next number of tens then becomes one of the factors of the product term. In principle, however, both neighboring tens can be used for the calculation, see the example of the square calculation of natural numbers between 40 and 50 (see Fig. 7.1).

| | |
|---|---|
| $41^2 = 40 \cdot 42 + 1^2 = 1680 + 1 = 1681$ | $41^2 = 32 \cdot 50 + 9^2 = 1600 + 81 = 1681$ |
| $42^2 = 40 \cdot 44 + 2^2 = 1760 + 4 = 1764$ | $42^2 = 34 \cdot 50 + 8^2 = 1700 + 64 = 1764$ |
| $43^2 = 40 \cdot 46 + 3^2 = 1840 + 9 = 1849$ | $43^2 = 36 \cdot 50 + 7^2 = 1800 + 49 = 1849$ |
| $44^2 = 40 \cdot 48 + 4^2 = 1920 + 16 = 1936$ | $44^2 = 38 \cdot 50 + 6^2 = 1900 + 36 = 1936$ |

$$45^2 = 40 \cdot 50 + 5^2 = 2000 + 25 = 2025$$

| | |
|---|---|
| $46^2 = 40 \cdot 52 + 6^2 = 2080 + 36 = 2116$ | $46^2 = 42 \cdot 50 + 4^2 = 2100 + 16 = 2116$ |
| $47^2 = 40 \cdot 54 + 7^2 = 2160 + 49 = 2209$ | $47^2 = 44 \cdot 50 + 3^2 = 2200 + 9 = 2209$ |
| $48^2 = 40 \cdot 56 + 8^2 = 2240 + 64 = 2304$ | $48^2 = 46 \cdot 50 + 2^2 = 2300 + 4 = 2304$ |
| $49^2 = 40 \cdot 58 + 9^2 = 2320 + 81 = 2401$ | $49^2 = 48 \cdot 50 + 1^2 = 2400 + 1 = 2401$ |

**Fig. 7.1**  Calculating the squares of natural numbers between 40 and 50

### 7.1.4  Calculating with Square Numbers: Checking the Final Digits

In order to avoid errors in mental arithmetic, consider a check using the final digits. A simple rule applies here (see Fig. 7.2):

---

**Rule 4**

For all natural numbers $k$ with $0 \leq k \leq 25$ applies:

The numbers $k^2$, $(50 - k)^2$, $(50 + k)^2$, and $(100 - k)^2$ match in the last two digits.

Therefore, it is sufficient for this check if you know the first 25 square numbers by heart.

The proof of this rule can be done with the help of binomial formulas.

The terms $(50 - k)^2$, $(50 + k)^2$, and $(100 - k)^2$ can be represented as the sum of $k^2$ and a product term:

$$(50 - k)^2 = 50^2 - 2 \cdot 50 \cdot k + k^2 = 2,500 - 100 \cdot k + k^2 = 100 \cdot (25 - k) + k^2$$
$$(50 + k)^2 = 50^2 + 2 \cdot 50 \cdot k + k^2 = 2,500 + 100 \cdot k + k^2 = 100 \cdot (25 + k) + k^2$$
$$(100 - k)^2 = 100^2 - 2 \cdot 100 \cdot k + k^2 = 10,000 - 200 \cdot k + k^2 = 100 \cdot (100 - 2 \cdot k) + k^2$$

Since one of the factors of the first summand is 100, the two final digits of this product are zero, and therefore the last two digits of the entire term depend only on the last two digits of $k^2$. ◄

| $n$ | $n^2$ | $n$ | $n^2$ | $n$ | $n^2$ | $n$ | $n^2$ |
|-----|-------|-----|-------|-----|-------|-----|-------|
| 00 | 00 | 50 | 2500 | 50 | 2500 | 100 | 10000 |
| 01 | 01 | 49 | 2401 | 51 | 2601 | 99 | 9801 |
| 02 | 04 | 48 | 2304 | 52 | 2704 | 98 | 9604 |
| 03 | 09 | 47 | 2209 | 53 | 2809 | 97 | 9409 |
| 04 | 16 | 46 | 2116 | 54 | 2916 | 96 | 9216 |
| 05 | 25 | 45 | 2025 | 55 | 3025 | 95 | 9025 |
| 06 | 36 | 44 | 1936 | 56 | 3136 | 94 | 8836 |
| 07 | 49 | 43 | 1849 | 57 | 3249 | 93 | 8649 |
| 08 | 64 | 42 | 1764 | 58 | 3364 | 92 | 8464 |
| 09 | 81 | 41 | 1681 | 59 | 3481 | 91 | 8281 |
| 10 | 100 | 40 | 1600 | 60 | 3600 | 90 | 8100 |
| 11 | 121 | 39 | 1521 | 61 | 3721 | 89 | 7921 |
| 12 | 144 | 38 | 1444 | 62 | 3844 | 88 | 7744 |
| 13 | 169 | 37 | 1369 | 63 | 3969 | 87 | 7569 |
| 14 | 196 | 36 | 1296 | 64 | 4096 | 86 | 7396 |
| 15 | 225 | 35 | 1225 | 65 | 4225 | 85 | 7225 |
| 16 | 256 | 34 | 1156 | 66 | 4356 | 84 | 7056 |
| 17 | 289 | 33 | 1089 | 67 | 4489 | 83 | 6889 |
| 18 | 324 | 32 | 1024 | 68 | 4624 | 82 | 6724 |
| 19 | 361 | 31 | 961 | 69 | 4761 | 81 | 6561 |
| 20 | 400 | 30 | 900 | 70 | 4900 | 80 | 6400 |
| 21 | 441 | 29 | 841 | 71 | 5041 | 79 | 6241 |
| 22 | 484 | 28 | 784 | 72 | 5184 | 78 | 6084 |
| 23 | 529 | 27 | 729 | 73 | 5329 | 77 | 5929 |
| 24 | 576 | 26 | 676 | 74 | 5476 | 76 | 5776 |
| 25 | 625 | 25 | 625 | 75 | 5625 | 75 | 5625 |

**Fig. 7.2** Squares of natural numbers between 0 and 100

### 7.1.5   Calculating with Square Numbers: Comparison of Methods

In the previous sections, various methods were presented which can be used for the mental calculation of square numbers.

In the following examples another method (method 3) is presented, in which the 1st binomial formula is applied.

**Example 1: Calculation of $37^2$**
- **Check the final digits:** $37^2$ has the same final digits as $13^2 = 1\underline{69}$.
- **Method 1 (stepwise calculation)**

$$\text{Auxiliary size: } 35^2 = 30 \cdot 40 + 5^2 = 1225$$
$$37^2 = 35^2 + (35 + 36) + (36 + 37) = 35^2 + 4 \cdot 36$$
$$= 1225 + 144 = 1369$$

This could also be illustrated as follows:
To complete the $35 \times 35$ square to a $37 \times 37$ square, you have to add $35 + 36 + 36 + 37$ unit squares.
- **Method 2 (calculation using equidistant numbers – a number of tens and a symmetrical partner)**

**Alternative 1:**

$$37^2 = (40 - 3) \cdot (34 + 3) = 40 \cdot 34 + 3^2 = 1360 + 9 = 1369$$

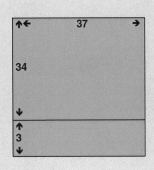 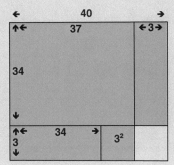

**Alternative 2:**

$$37^2 = (30 + 7) \cdot (44 - 7) + 7^2 = 30 \cdot 44 + 7^2 = 1320 + 49 = 1369$$

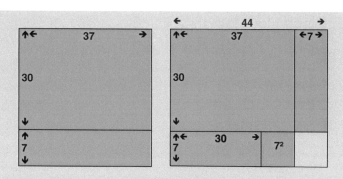

• **Method 3 (applying the 1st binomial formula)**

$$37^2 = (30+7)^2 = 30^2 + 2 \cdot 30 \cdot 7 + 7^2 = 30 \cdot (30 + 2 \cdot 7) + 7^2$$
$$= 30 \cdot 44 + 7^2 = 1320 + 49 = 1369$$

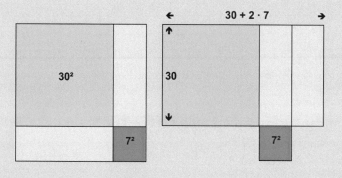

**Suggestions for Reflection and for Investigations**
**A 7.1** Calculate the square of 58 [of 84, of 73] using the three methods.

## 7.2   Number Cycles

If you only consider the last digit when squaring a natural number, you get the following transitions:

$$1 \to 1; \ 2 \to 4; \ 3 \to 9; \ 4 \to 6; \ 5 \to 5; \ 6 \to 6; \ 7 \to 9; \ 8 \to 4; \ 9 \to 1; \ 0 \to 0$$

The final digits 1, 5, 6, and 0 lead to themselves. The other final digits are contained in **number chains**:

$$2 \to 4 \to 6 \to 6; \ 3 \to 9 \to 1 \to 1; \ 7 \to 9 \to 1 \to 1; \ 8 \to 4 \to 6 \to 6$$

So if you start with a natural number whose last digit is 2, then this chain ends at a number with last digit 6 and so also does a chain starting with the digit 8. Chains starting with the final digit 3 or the final digit 7 also end up at the same number.

The investigation of the last two digits of a natural number is still more exciting. Here one can discover remarkable **number cycles** when squaring the numbers continuously.

---

**Example: A Number Cycle Resulting From the Start Number 17**

$$17 \to 89 \to 21 \to 41 \to 81 \to 61 \to 21 \to 41 \to \ldots$$

If you square any number that ends in 17, the square number of that number always ends in 89, the square of that number always ends in 21, its square ends in 41, its square ends in 81, its square ends in 61, its square ends in 21.

Then the sequence $21 - 41 - 81 - 61$ is repeated infinitely often; the cycle has the length 4.

---

In the following, we systematically examine all initial numbers with final digits between 00 and 99.

## 7.2.1   Number Cycles Ending After One or Two Steps

The squares of natural numbers whose final digit is 0 have "00" as their final digits.
So these cycles all end at "00".

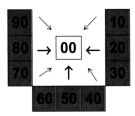

The squares of natural numbers whose final digit is 5 have "25" as final digits.
So these cycles all end at "25".

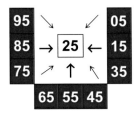

The squares of 07, 43, 57, and 93 have "49" as final digits. The squares of 01, 49, 51, and 99 have "01" as final digits.

So these cycles all end at "01".

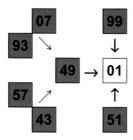

The squares of 18, 32, 68, and 82 have "24" as final digits. The squares of 24, 26, 74, and 76 have "76" as final digits.

So these cycles all end at "76".

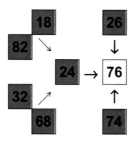

## 7.2.2   Periodic Cycles

The following number cycles end in a periodic cycle of length 4:

- The squares of 02, 48, 52, and 98 have "04" as final digits; the squares of 04, 46, 54, and 96 have "16" as final digits.
- The squares of 12, 38, 62, and 88 have "44" as final digits; the squares of 06, 44, 56, and 94 have "36" as final digits.
- The squares of 22, 28, 72, and 78 have "84" as final digits; the squares of 16, 34, 66, and 84 have "56" as final digits.
- The squares of 08, 42, 58 and 92 have "64" as final digits; the squares of 14, 36, 64, and 86 have "96" as final digits.

The squares of 16, 56, 36, and 96 form a periodic cycle of length 4 (see Fig. 7.3a).

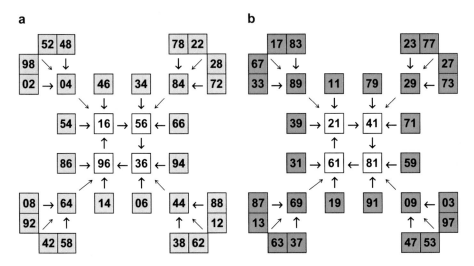

Fig. 7.3  **a, b** Squaring natural numbers: the two periodic cycles of length 4

- The squares of 17, 33, 67, and 83 have "89" as final digits; the squares of 11, 39, 61, and 89 have "21" as final digits.
- The squares of 23, 27, 73, and 77 have "29" as final digits; the squares of 21, 29, 71, and 79 have "41" as final digits.
- The squares of 13, 37, 53, and 87 have "69" as final digits; the squares of 19, 31, 69, and 81 have "61" as final digits.
- The squares of 03, 47, 53, and 97 have "09" as final digits; the squares of 09, 41, 59, and 91 have "81" as final digits.

The squares of 21, 41, 81, and 61 form a periodic cycle of length 4 (see Fig. 7.3b).

In Sect. 7.1.3 we have learnt that four two-digit numbers, which complement each other to 50 or 100 respectively, have the same two final digits when squared; this property is also comprised in the two figures of Fig. 7.3.

## 7.3    Number Cycles Modulo *n*

In the seventeenth century mathematicians were also concerned with the rules of a special arithmetic, the calculation with so-called congruencies. Leonhard Euler (1707–1783) and Christian Goldbach (1690–1764) had an intensive correspondence on this topic. Carl Friedrich Gauss (1777–1855) developed a theory of congruencies in his *Disquisitiones Arithmeticae* published in the year 1801.

In principle, it is a matter of calculating with the remainders that occur when you perform a division. In Sect. 7.2, we examined which final digits occur when squaring a natural number, and then what the last two digits are. This is a calculation of congruencies modulo 100, because only the last two digits of a natural number are considered.

From our everyday life we know the calculation of congruences modulo 24, because the day has 24 h. Since our watches usually only show 12 h, we calculate modulo 12.

For example, if we want to calculate what time it will be in 77 h, then we add the 77 to the current time, divide the sum by 24 and are only interested in what remains after division by 24. In practice, instead of dividing by 24, you might prefer to subtract the number 24 repeatedly until you have a remainder smaller than 24, but this is exactly the method used to introduce division in primary school.

---

**Example: Calculating modulo 24**

The current time is 10 o'clock; in 77 h it will be

$$10 + 77 (mod\ 24) \equiv 87\ (mod\ 24) \equiv 15\ (mod\ 24),$$

so it's 3:00 pm.

---

The notation "$\equiv$" for the *equality of remainders with regard to the same divisor* was introduced by C. F. Gauss. You read this as "87 is *congruent to* 15 *modulo* 24".

If one examines the remainders modulo 24 after squaring natural numbers, then one notices that this is rather boring: understandably, only the square numbers 0, 1, 4, 9, and 16 appear as remainders, in addition to the number 12 (see following table).

More extensive cycles do not occur – a chain ends at the latest after the second squaring,

for example, $10 \rightarrow 4 \rightarrow 16 \rightarrow 16$.

What is striking here, however, is the symmetry of the relationships.

| *n* | 0 | 1 | 2 | 3 | 4 | 5 | 6 | 7 | 8 | 9 | 10 | 11 |
|-----|---|---|---|---|----|---|----|---|----|---|----|----|
| $n^2$ | 0 | 1 | 4 | 9 | 16 | 1 | 12 | 1 | 16 | 9 | 4 | 1 |
| *n* | 12 | 13 | 14 | 15 | 16 | 17 | 18 | 19 | 20 | 21 | 22 | 23 |
| $n^2$ | 0 | 1 | 4 | 9 | 16 | 1 | 12 | 1 | 16 | 9 | 4 | 1 |

However, if you change the number which is chosen to calculate the remainder, you will find interesting proporties.

For example, with regard to the congruencies modulo 11, one finds the structure shown in the following figure:

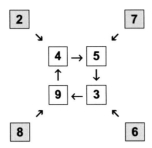

**Suggestions for reflection and for investigations**

**A 7.2** Examine, which cycles occur when squaring modulo $n$.

1. A cycle of length 6 for $n = 19$;
2. A cycle of length 20 for $n = 23$;
3. A complex cycle of length 4 for $n = 25$;
4. A cycle of length 6 for $n = 27$.

*Note:* The use of a spreadsheet is recommended for the systematic investigation of cycles modulo $n$. In Excel®, the command $= MOD(number; divisor)$ can be used for this purpose.

## 7.4    Number Cycles for Higher Powers

In the following, suggestions are made to research the properties of cube numbers ($3^{rd}$ power). Various properties can be read from the following graphics. To prove the properties, however, one needs knowledge of the binomial formulas for $3^{rd}$ powers.

### 7.4.1    Analyzing the Last Two Final Digits of Cubic Numbers

If you consider the last digits of natural numbers and determine the last digits of the corresponding cube numbers, you will discover some interesting properties for the numbers with the final digits from 1 to 4 and from 6 to 9.

You may consider by yourself which properties apply if the final digit is 0 or 5.

- The $3^{rd}$ power of the numbers with the final digits "01" or "99" or "49" or "51" end exactly with these final digits.

- The 3$^{rd}$ power of 26 has the final digits "76"; the 3$^{rd}$ power of 76 has the final digits "76". The 3$^{rd}$ power of 74 has the final digits "24"; the 3$^{rd}$ power of 24 has the final digits "24".
- The 3$^{rd}$ power of 07 has the final digits "43"; the 3$^{rd}$ power of 43 has the final digits "07". The 3$^{rd}$ power of 93 has the final digits "57"; the 3$^{rd}$ power of 57 has the final digits "93".
- The 3$^{rd}$ power of 18 has the final digits "32"; the 3$^{rd}$ power of 32 has the final digits "68". The 3$^{rd}$ power of 82 has the final digits "68"; the 3$^{rd}$ power of 68 has the final digits "32".

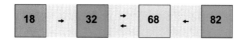

- The 3$^{rd}$ power of 06 has the final digits "16"; the 3$^{rd}$ power of 16 has the final digits "96".

  The 3$^{rd}$ power of 66 has the final digits "96"; the 3$^{rd}$ power of 96 has the final digits "36".

  The 3$^{rd}$ power of 46 has the final digits "36"; the 3$^{rd}$ power of 36 has the final digits "56".

  The 3$^{rd}$ power of 86 has the final digits "56"; the 3$^{rd}$ power of 56 has the final digits "16".

  The numbers with the final digits "16", "96", "36", and "56" form a periodic cycle of length 4.

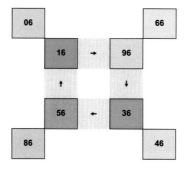

- The numbers with the final digits "03", "27", "83", and "87" form a periodic cycle of length 4.

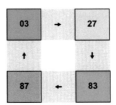

**Suggestions for Reflection and for Investigations**
A 7.3 Examine which other periodic number cycles occur.

## 7.4.2  Investigations of the Last Three Final Digits of a Cubic Number

Of course, it is much more interesting to consider not only the last *two* digits of the third power of natural numbers, but even the last *three* digits of a number and the three final digits of the corresponding cubic numbers. All natural numbers with the final digit 0 are not interesting, because their $3^{rd}$ powers have the final digits 000.

In the suggestions given below, different types of cycles are listed, of which some examples are given.

**Suggestions for Reflection and for Investigations**
**A 7.4**

1. The cycles of numbers whose final digit is 5 are not very interesting. Show: The natural numbers whose final digit is 5 lead in the $3^{rd}$ power to numbers with the final digits 125, 875, 375, or 625.
2. There are also numbers whose three final digits always trace back to themselves, for example, 001 and 999. Consider which other (six) natural numbers in the $3^{rd}$ power trace back to themselves.
3. There are partner numbers that lead reciprocally to each other, for example, 307 and 193. Consider which other (two) pairs of natural numbers lead reciprocally to each other in the $3^{rd}$ power.
4. The numbers with the final digits 124, 374, and 874 lead in the $3^{rd}$ power to a number with the final digits 624. Which other numbers have an analogous property as the numbers 124, 374, and 874?

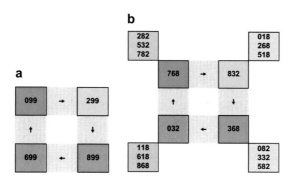

Fig. 7.4  **a, b** Examples of "cycles of length 4" (cubes modulo 1000)

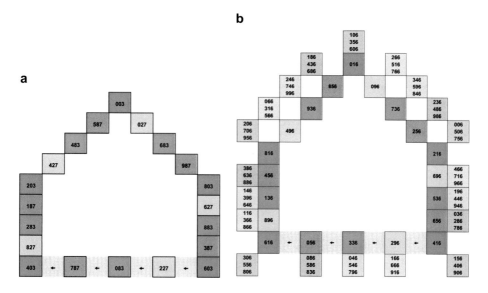

Fig. 7.5  **a, b** Examples of "cycles of length 20" (cubes modulo 1000)

5. There is a special feature with the numbers with the final digits 068, 318, and 818, which lead in the 3rd power to the final digits 432. And there are partner numbers, too. Which are these?
6. The remaining natural numbers form a cycle of length 4 (see Fig. 7.4a, b) or even a cycle of length 20 (see Fig. 7.5a, b).

## 7.5     References to Further Literature

On **Wikipedia** you can find further information and literature on the keywords in English (German, French):

- Mental calculation (Kopfrechnen, calcul mental)
- Modulo operation (Kongruenz, Congruence sur les entiers/Modulo (opération))

Extensive informations can be found at **Wolfram Mathworld** under the keyword:

- Congruence

Among the many stimulating contributions in

- Erickson, Martin, *Beautiful Mathematics,* The Mathematical Association of America Inc., 2011

you will find a cycle modulo 25 as well as further information on this topic.

# Partitions of Regular Polygons

**8**

*The whole is greater than the sum of its parts.*

*(Aristotle, Greek philosopher, 384–322 B.C.)*

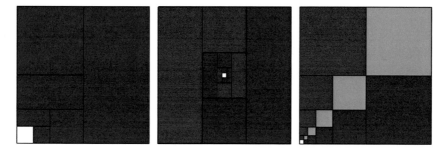

A regular polygon, for example, a square, can be dissected in different ways by continued partition.

## 8.1 Continued Bisection

In the following illustrations, you find one of the many ways of dividing up a square: First divide the square into two congruent rectangles. One of the two rectangles is colored blue, the other remains white. Then the white rectangle is divided into two congruent squares. One of the two squares is colored blue, the other remains white.

© Springer-Verlag GmbH Germany, part of Springer Nature 2021
H. K. Strick, *Mathematics is Beautiful*, https://doi.org/10.1007/978-3-662-62689-4_8

After these two steps, the situation for the (still) uncolored white square is similar to the
situation at the beginning. In the next steps, this white square is first bisected lengthwise
again and one half is colored blue, then the white rectangle is bisected crosswise and one
of the squares is colored blue.

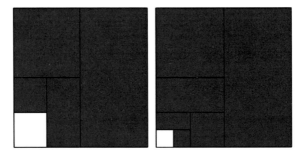

If you continue to divide the white square into two rectangles and divide one of the two
rectangles into two squares, and color each time one of the two subareas created by division, the uncolored area of the initial square becomes smaller and smaller. If the process
is carried out infinitely often, the entire square is colored.

The initial square is dissected by the continued process of bisection into a sequence of
rectangles and squares, each half the size of its predecessor.

The following, therefore, applies to the area of the square and its subareas:

$$\frac{1}{2} + \frac{1}{2} \cdot \frac{1}{2} + \frac{1}{2} \cdot \frac{1}{2} \cdot \frac{1}{2} + \frac{1}{2} \cdot \frac{1}{2} \cdot \frac{1}{2} \cdot \frac{1}{2} + \frac{1}{2} \cdot \frac{1}{2} \cdot \frac{1}{2} \cdot \frac{1}{2} \cdot \frac{1}{2} + \ldots =$$
$$\frac{1}{2} + \frac{1}{4} + \frac{1}{8} + \frac{1}{16} + \frac{1}{32} + \ldots = \frac{1}{2^1} + \frac{1}{2^2} + \frac{1}{2^3} + \frac{1}{2^4} + \frac{1}{2^5} + \ldots = 1$$

That is:

$$\frac{1}{2} + \frac{1}{4} + \frac{1}{8} + \frac{1}{16} + \frac{1}{32} + \ldots = \frac{1}{2^1} + \frac{1}{2^2} + \frac{1}{2^3} + \frac{1}{2^4} + \frac{1}{2^5} + \ldots = 1 \qquad (8.1)$$

A square can also be bisected by using a diagonal: This method of bisection creates two
congruent isosceles right-angled triangles. One of the resulting triangles is colored and
the other is bisected with the help of the altitude (axis of symmetry) in the next step. And

this procedure can be continued infinitely. In contrast to the first method of dissection, where we had squares and rectangles, the isosceles right-angled triangle always has to be bisected in the same way (from step 2 onward).

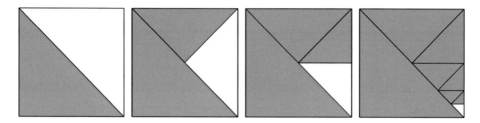

## 8.2  Continued Trisection

A square can be dissected into two subareas of equal size by continued trisections:

First the initial square is divided into three congruent rectangles. One of the outer rectangles is colored blue, the other red; the rectangle in the middle remains white. Second, the central rectangle can be divided into three squares of equal size. The upper one of the squares is colored red, the lower one blue, and the square in the middle remains white.

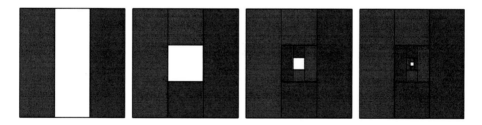

After these two steps, the situation for the uncolored white square is the same as at the beginning. In the next steps this white square is again divided into three parts by vertical lines, this time the left third is colored blue, the right third red, then the white rectangle is divided into three congruent parts by horizontal lines, this time colored red at the bottom and blue at the top.

If you continue to divide the white square and the rectangle respectively and color the areas created by division, the uncolored part of the initial square becomes smaller and smaller. If the process is carried out infinitely often, the entire initial square is colored with two colors.

The fact that the colors in the left and right rectangles were swapped in the odd steps to the colors in the top and bottom squares in the even steps results in a spiral decomposition of the original square.

For the area of the red or blue colored parts of the square and its parts, the following applies:

$$\frac{1}{3}+\frac{1}{3}\cdot\frac{1}{3}+\frac{1}{3}\cdot\frac{1}{3}\cdot\frac{1}{3}+\frac{1}{3}\cdot\frac{1}{3}\cdot\frac{1}{3}\cdot\frac{1}{3}+\frac{1}{3}\cdot\frac{1}{3}\cdot\frac{1}{3}\cdot\frac{1}{3}\cdot\frac{1}{3}+\ldots=\frac{1}{3}+\frac{1}{9}+\frac{1}{27}+\frac{1}{81}+\frac{1}{243}+\ldots=\frac{1}{3^1}+\frac{1}{3^2}+\frac{1}{3^3}+\frac{1}{3^4}+\frac{1}{3^5}+\ldots=\frac{1}{2}$$

That is:

$$\frac{1}{3}+\frac{1}{9}+\frac{1}{27}+\frac{1}{81}+\frac{1}{243}+\ldots=\frac{1}{3^1}+\frac{1}{3^2}+\frac{1}{3^3}+\frac{1}{3^4}+\frac{1}{3^5}+\ldots=\frac{1}{2} \qquad (8.2)$$

**Suggestions for Reflection and for Investigations**

**A 8.1**: What figure would result if the coloring in the 3rd and 4th step had been done as in the 1st and 2nd step (i.e., without exchanging left and right or up and down)?

**A 8.2**: How should the square be trisected so that after an infinite number of steps the left half of the initial square is colored red and the right half blue?

**A 8.3**: Analyze the following figure: In which way the square and the isosceles right-angled triangles are divided and colored? What is the ratio of the yellow-colored areas?

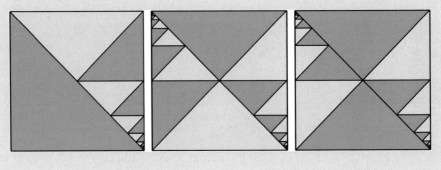

Instead of a square, you can also dissect an equilateral triangle of unit area:

The equilateral triangle is divided into three congruent parts by connecting the center of the triangle with the vertices. Two parts are colored, one part of the triangle remains white.

In the next step, the white triangle is dissected into three triangles of equal size. This is done by dividing the base into three segments. Since the altitudes of the resulting triangles are of equal length, the triangles have the same area. Two of the three triangles are colored, one remains white.

Since this uncolored triangle is again an equilateral triangle, steps 1 and 2 can be applied to this triangle again. If one carries out the process infinitely often, the initial triangle is colored in two colors, thus confirming formula (8.2).

**Suggestions for Reflection and for Investigations**
**A 8.4**: What is the side length of an equilateral triangle of unit area?

## 8.3 Continued Quadrisection

The initial square is divided into four subsquares of equal size. Three of the squares are colored red, green, and blue, one square remains white. This white square is then divided again in the 2nd step into four equally sized smaller squares. Three of the squares are colored red, green, and blue, one square remains white, and so on.

If you continue to divide the white square into four congruent squares and color three of them, the uncolored area of the initial square becomes smaller and smaller. If one carries out the process infinitely often, the entire square is colored with three colors.

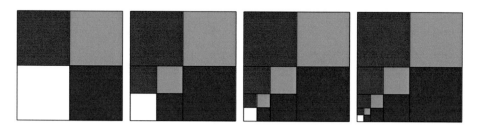

Therefore, the following applies for example to the red colored areas of the square (to the blue or green colored similarly):

$$\frac{1}{4}+\frac{1}{4}\cdot\frac{1}{4}+\frac{1}{4}\cdot\frac{1}{4}\cdot\frac{1}{4}+\frac{1}{4}\cdot\frac{1}{4}\cdot\frac{1}{4}\cdot\frac{1}{4}+\frac{1}{4}\cdot\frac{1}{4}\cdot\frac{1}{4}\cdot\frac{1}{4}\cdot\frac{1}{4}+\ldots=\frac{1}{4}+\frac{1}{16}+\frac{1}{64}+\frac{1}{256}+\frac{1}{1,024}+\ldots=\frac{1}{4^1}+\frac{1}{4^2}+\frac{1}{4^3}+\frac{1}{4^4}+\frac{1}{4^5}+\ldots=\frac{1}{3}$$

That is:

$$\frac{1}{4}+\frac{1}{16}+\frac{1}{64}+\frac{1}{256}+\frac{1}{1,024}+\ldots=\frac{1}{4^1}+\frac{1}{4^2}+\frac{1}{4^3}+\frac{1}{4^4}+\frac{1}{4^5}+\ldots=\frac{1}{3} \quad (8.3)$$

**Suggestions for Reflection and for Investigations**

**A 8.5**: In the following figures a possible quadrisection of a square is suggested. How could the isosceles triangles be colored so that the validity of the formula (8.3) is illustrated here as well?

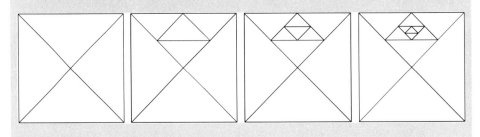

Instead of a square of 1 square unit, an equilateral triangle with this area can be chosen as an initial figure (see the following figures).

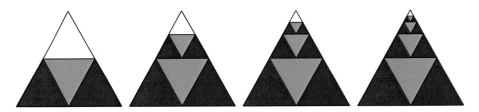

**A 8.6**: Describe the process of quadrisection, shown in the following figures in more detail. In particular specify the lengths of the sides of the occurring trapezoids. What is special about the side lengths of the trapezoids?

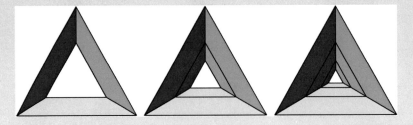

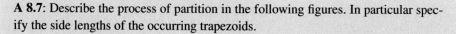

**A 8.7**: Describe the process of partition in the following figures. In particular specify the side lengths of the occurring trapezoids.

## 8.4    Continued Dissection into Five Equal Parts

In the following figures you see one of the many ways to dissect the initial square into five equal parts and color four of them with different colors. In the next step, the uncolored square is divided again into five parts of equal size and four of them are colored in such a way that a white square remains.

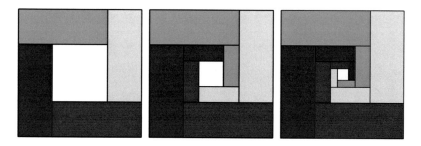

If you continue to divide the white square accordingly, the noncolored area of the initial square becomes smaller and smaller. If one carries out the process infinitely, the entire square is colored with four colors.

So for the area of the red or blue or green or yellow colored parts of the square and its subareas applies:

$$\frac{1}{5} + \frac{1}{5} \cdot \frac{1}{5} + \frac{1}{5} \cdot \frac{1}{5} \cdot \frac{1}{5} + \frac{1}{5} \cdot \frac{1}{5} \cdot \frac{1}{5} \cdot \frac{1}{5} + \frac{1}{5} \cdot \frac{1}{5} \cdot \frac{1}{5} \cdot \frac{1}{5} \cdot \frac{1}{5} + \ldots$$

$$= \frac{1}{5} + \frac{1}{25} + \frac{1}{125} + \frac{1}{625} + \frac{1}{3,025} + \ldots = \frac{1}{5^1} + \frac{1}{5^2} + \frac{1}{5^3} + \frac{1}{5^4} + \frac{1}{5^5} + \ldots = \frac{1}{4}$$

That is:

$$\frac{1}{5} + \frac{1}{25} + \frac{1}{125} + \frac{1}{625} + \frac{1}{3,025} + \ldots = \frac{1}{5^1} + \frac{1}{5^2} + \frac{1}{5^3} + \frac{1}{5^4} + \frac{1}{5^5} + \ldots = \frac{1}{4} \quad (8.4)$$

**Suggestions for Reflection and for Investigations**

**A 8.8**: Describe the process of dividing into five parts shown in the figures above in more detail. Specify the side lengths of the rectangles that occur.

**A 8.9**: Describe the process of dissection shown in the following figures. Specify the side lengths of the rectangles that occur.

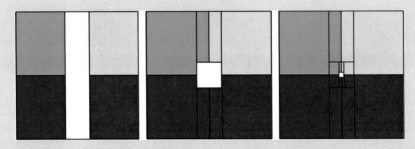

**A 8.10**: Describe the following process of dividing a square in five parts in more detail. Specify the side lengths of the occurring trapezoids.

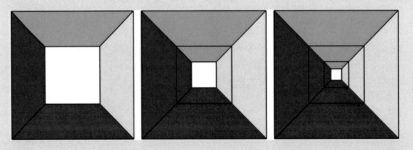

**A 8.11**: Describe the process of dissection of a square as shown in the following figures. Specify in particular the side lengths of the pentagons.

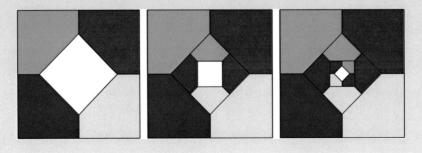

## 8.5   Continued Dissections into $n$ Subareas of Equal Size

A formula for the sum of the powers of $\frac{1}{6}$ can be found analogously, that is

$$\frac{1}{6^1} + \frac{1}{6^2} + \frac{1}{6^3} + \frac{1}{6^4} + \frac{1}{6^5} + \ldots = \frac{1}{5}. \tag{8.5}$$

This can be illustrated by the continued dissection of the the area of a regular pentagon.

**1<sup>st</sup> Possibility**:

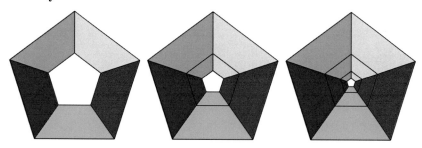

**2<sup>nd</sup> Possibility**:

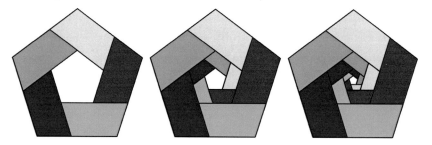

The ideas underlying the two methods of dissection can be transferred to any regular $n$-sided figure.

The first idea corresponds to the decomposition of a regular polygon into axially symmetrical triangles, whose bases are also the edges of the polygon; the center of the $n$-sided figure is the apex for all triangels. Each of the axially symmetrical triangles then is composed of an infinite "sum" of symmetrical trapezoids.

The other idea leads to a spiral-like decomposition of the regular $n$-sided figure. In the center of the polygon, a smaller regular $n$-sided figure is drawn which is similar to the initial figure; the area of the polygon in the center corresponds exactly to the $(n + 1)$-th part of the total area. The remaining area consists of $n$ congruent trapezoids.

From both methods of partition we obtain the following general formula:

**Formula**

**Infinite sum with limit $\frac{1}{n}$**

For natural numbers $n$ with $n \geq 1$ applies:

$$\frac{1}{(n+1)^1} + \frac{1}{(n+1)^2} + \frac{1}{(n+1)^3} + \ldots = \frac{1}{n} \qquad (8.6)$$

◄

## 8.6   Geometric Sequences and Series

Formula (8.6) is a special case of a formula for the so-called geometric series.

**Definition**

**Geometric sequence and geometric series**

A sequence of numbers $(a_n)_{n \in \mathbb{N}}$ is called a geometric sequence or geometric progression, if the quotient $\frac{a_{n+1}}{a_n}$ of two consecutive elements of the sequence is constant; this constant is usually designated as $r$ (= ratio).

The sum of the elements of a geometric sequence $(a_n)_{n \in \mathbb{N}}$ is referred to as **geometric series** $(s_n)_{n \in \mathbb{N}}$ with $s_n = a_1 + a_2 + \ldots + a_n$. ◄

The elements of a geometric sequence can be calculated recursively for $n > 1$ according to the rule $a_{n+1} = a_n \cdot r$ (according to definition) and explicitly using $a_n = a_1 \cdot r^{n-1}$, where $a_1$ is the initial element in the sequence.

The following applies to the corresponding geometric series:

**Theorem**

**Calculation of the elements of a geometric series**

A geometric sequence with initial element $a_1$ and constant factor $r$ is given.

Then the following applies to the elements of the geometric series:

$$s_n = a_1 \cdot \frac{1 - r^n}{1 - r} = a_1 \cdot \frac{r^n - 1}{r - 1} \qquad (8.7)$$

For $r < 1$ the geometric series converges toward the limit $s$:

$$s = \lim_{n \to \infty} (s_n) = a_1 \cdot \frac{1}{1 - r} \qquad (8.8)$$

◄

The proof of the formula (8.7) can be obtained by multiplying both sides of the equation by the factor $(1 - r)$ (see for example, the article on Wikipedia).

The validity of the formula (8.8) for the limit of the geometric series with the initial element $a = a_1$ results from the following figure.

Here the triangles $ABC$ and $ECD$ are similar to each other, so the following applies: $AB : AC = CE : DE$, therefore:

$$(a + ar + ar^2 + ar^3 + \ldots) : 1 = a : (1 - r)$$

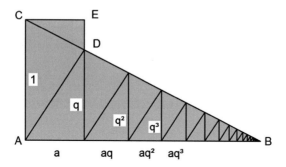

In the previous sections, specific geometric sequences were examined, namely
in Sect. 8.1 with $a_1 = \frac{1}{2}$ and $r = \frac{1}{2}$ in Sect. 8.2 with $a_1 = \frac{1}{3}$ and $r = \frac{1}{3}$,
in Sect. 8.3 with $a_1 = \frac{1}{4}$ and $r = \frac{1}{4}$ in Sect. 8.4 with $a_1 = \frac{1}{5}$ and $r = \frac{1}{5}$,
in Sect. 8.5 with $a_1 = \frac{1}{6}$ and $r = \frac{1}{6}$ in general $a_1 = \frac{1}{n+1}$ and $r = \frac{1}{n+1}$.
The limit $s$ of these geometric series was obtained using the chosen graphical method.
A formal calculation would be more complicated here:

$$s = \frac{1}{n+1} \cdot \frac{1}{1 - \frac{1}{n+1}} = \frac{1}{n+1} \cdot \frac{1}{\frac{n+1}{n+1} - \frac{1}{n+1}} = \frac{1}{n+1} \cdot \frac{1}{\frac{n}{n+1}} = \frac{1}{n}$$

**Suggestions for Reflection and for Investigations**

**A 8.12**: Compare the sequence of area dissections of the regular 8-sided figure in the first three of the following figures with those of the last three figures. In the first row the radius is continuously halved, in the second row it is continuously divided by 3.

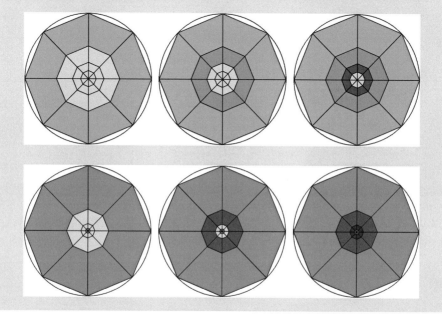

## 8.7    Dissection of Regular Polygons into Subareas of Equal Size

According to a theorem of similarity, the ratio between the areas of similar geometric figures is equal to the squares of the ratio of corresponding lengths of those figures.

Therefore, it is actually not a very challenging problem if it is only a matter of dividing the area of a regular $n$-sided figure into $k = 2, 3, \ldots$ parts of equal size: You only have to multiply the side length of the basic figure with $\frac{1}{\sqrt{k}}$ to get the figure you want.

When bisecting a figure into two equally sized areas, the side length of the initial figure must therefore be multiplied by $\frac{1}{\sqrt{2}}$, when dividing into three subareas of equal size by $\frac{1}{\sqrt{3}}$ or with $\sqrt{\frac{2}{3}}$, when divided into four areas of equal size with $\frac{1}{\sqrt{4}} = \frac{1}{2}$ and $\sqrt{\frac{3}{4}}$ respectively and $\sqrt{\frac{2}{4}} = \frac{1}{\sqrt{2}}$ and so on.

The following figures show dissections of regular polygons, starting in a vertex of the polygon.

**Dissections with two equally sized subareas**

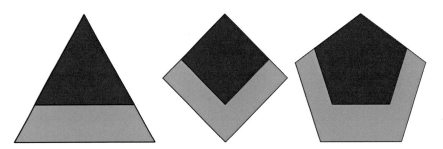

**Dissections with three equally sized subareas**

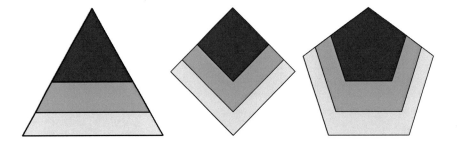

**Dissections with four equally sized subareas**

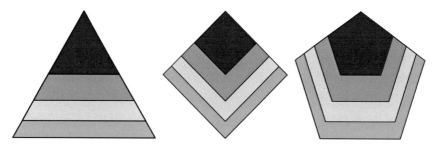

In the following figures the dissection of the figures starts in the center of the polygon.

**Dissections with two equally sized subareas**

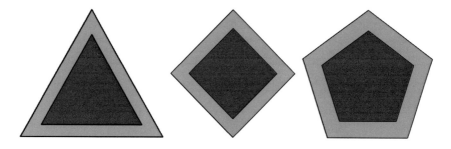

**Dissections with three equally sized subareas**

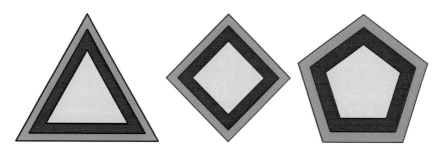

Concluding remark: As already stated in Chap. 4 on circular rings, one has no "feeling" at all whether the subareas in the figures are of equal size. This is especially true for the figures in the following illustrations, which are divided by horizontal lines.

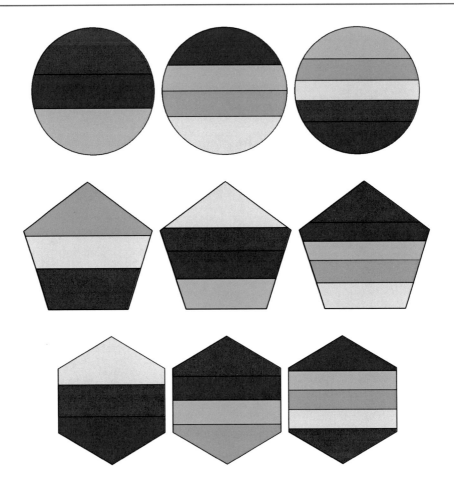

**Suggestions for Reflection and for Investigations**
**A 8.13**: In order to draw the last figures with three, four, five equally sized subareas, the lengths of the sides must be calculated. Which calculations are necessary?

## 8.8    References to Further Literature

On **Wikipedia** you can find further information and literature on the keywords in English (German, French):

- Geometric progression/sequence (Geometrische Folge, Suite géométrique)
- Geometric series (Geometrische Reihe, Série géométrique)

Extensive informations can be found at **Wolfram Mathworld** under the keywords:

- Geometric Sequence, Geometric Series

Many inspiring examples on the topic *Area divisions* can be found in the books of Roger B. Nelsen, see *general references.*

# Weighing in the Ternary Numeral System

**9**

*People who do not know anything about algebra cannot imagine the wonderful things that can be achieved using this science.*

*(Gottfried Wilhelm Leibniz, German mathematician and philosopher, 1646–1716)*

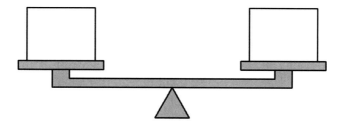

This chapter is initially only about the weighing of objects with the help of a beam balance – ultimately it is about the structure of the number system.

**A Weighing Problem**

Five special balance weights of 1, 3, 9, 27, and 81 g are given.

It is claimed that any object can be weighted with the balance weights and the beam balance to an accuracy of 1 g if the object's weight is not greater than 121 g.

© Springer-Verlag GmbH Germany, part of Springer Nature 2021
H. K. Strick, *Mathematics is Beautiful*, https://doi.org/10.1007/978-3-662-62689-4_9

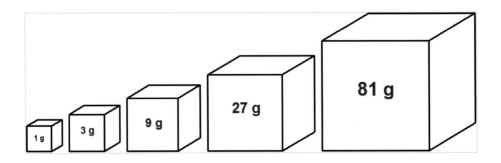

## 9.1 Solving the Simple Cases of the Weighing Problem

An object whose weight is known is placed on the left-hand scale pan. The aim is to investigate how the five different balance weights can be used to balance the scale.

**Example 1**
If the object weighs 10 g, put the balance weights of 1 and 9 g on the right scale pan.

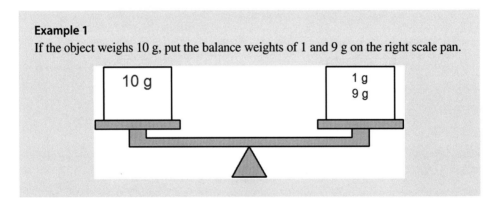

**Example 2**
If the object weighs 31 g, put the balance weights of 1, 3, and 27 g on the right scale pan.

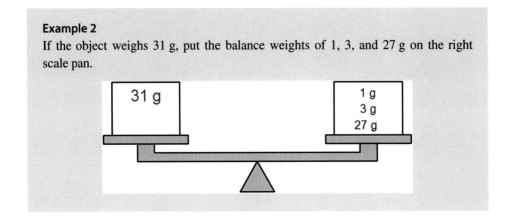

But only for a few weights of the object is the weighing as easy as in these two examples.

The object's weights of 10 or 31 g can be noted as a five-digit number consisting of zeros and ones, depending on whether a balance weight was used or not:

$$10g = \mathbf{0} \cdot 81g + \mathbf{0} \cdot 27g + \mathbf{1} \cdot 9g + \mathbf{0} \cdot 3g + \mathbf{1} \cdot 1g, \; in \; short : \mathbf{00101}$$

$$31g = \mathbf{0} \cdot 81g + \mathbf{1} \cdot 27g + \mathbf{0} \cdot 9g + \mathbf{1} \cdot 3g + \mathbf{1} \cdot 1g, \; in \; short : \mathbf{01011}$$

There are only $2 \cdot 2 \cdot 2 \cdot 2 \cdot 2 = 2^5 = 32$ numbers that have five digits and consist only of zeros and ones, including the number zero (= **00000**) that is, there are 31 object's weights that are easy to balance.

The largest object's weight that can be represented easily is 121 g; then each of the five balance weights is used:

$$121g = \mathbf{1} \cdot 81g + \mathbf{1} \cdot 27g + \mathbf{1} \cdot 9g + \mathbf{1} \cdot 3g + \mathbf{1} \cdot 1g = \mathbf{11111}$$

For the remaining $121 - 31 = 90$ object's weights, another weighing procedure must exist if the above claim is correct.

> **Suggestions for Reflection and for Investigations**
> **A 9.1**: State all 31 object's weights that can be represented easily.

## 9.2    Solution of the Other Cases of the Weighing Problem

The remaining 90 cases are easy to solve, too: Unlike the simple cases, different balance weights can also be placed on the left side of the weighing pan. This increases the weight of the object to be weighed. Then subtract these additional weights on the left side from the sum of the balance weights on the right side of the weighing pan.

> **Example 3**
> If the object weighs 8 g, put the 1 g balance weight on the left pan; then the 9 g balance weight is adequate to put on the right pan.
>
>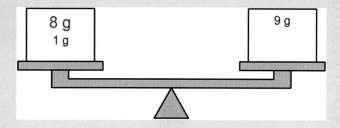

**Example 4**
If the object weighs 17 g, put the balance weights of 1 and 9 g on the left pan; the balance weight of 27 g is adequate to put on the right pan.

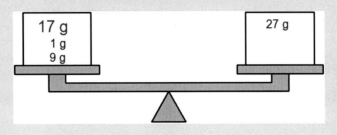

**Example 5**
If the object weighs 55 g, place the 27 g balance weight on the left pan, and the 1 g and 81 g balance weights on the right pan.

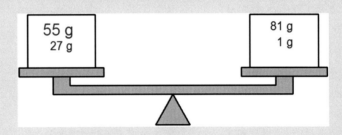

**Example 6**
If the object weighs 97 g, put the balance weights of 3 and 9 g on the left pan, and the balance weights of 1, 27, and 81 g on the right pan.

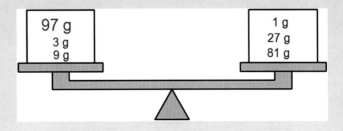

The additional balance weights required on the left pan are noted as above in the form of an equation; this means for example 3:

$$8g + 1 \cdot 1g = 0 \cdot 81g + 0 \cdot 27g + 1 \cdot 9g + 0 \cdot 3g + 0 \cdot 1g$$

This equation can be rearranged – as is usual with equations:

$$8g = 0 \cdot 81g + 0 \cdot 27g + 1 \cdot 9g + 0 \cdot 3g + (-1) \cdot 1g$$

So the condensed version of the equation is: $10g = 0010(-1)$.

In the other examples, this results in the following:

$$17g = 0 \cdot 81g + 1 \cdot 27g + (-1) \cdot 9g + 0 \cdot 3g + (-1) \cdot 1g = 01(-1)0(-1)$$
$$55g = 1 \cdot 81g + (-1) \cdot 27g + 0 \cdot 9g + 0 \cdot 3g + 1 \cdot 1g = 1(-1)001$$
$$97g = 1 \cdot 81g + 1 \cdot 27g + (-1) \cdot 9g + (-1) \cdot 3g + 1 \cdot 1g = 11(-1)(-1)1$$

For each of the 90 remaining cases, such a representation is possible using 0, 1, or $-1$. However, this will not be discussed here; rather, we will first examine another issue.

## 9.3   Representation of Natural Numbers in the Ternary Numeral System

What is striking about the selected weights of 1, 3, 9, 27, and 81 g is that they are the first powers of the number 3 $(3^0 = 1, \ 3^1 = 3, \ 3^2 = 9, \ 3^3 = 27, \ 3^4 = 81)$.

For the representation of numbers in the ternary numeral system, usually the **digits 0, 1, and 2 are used.**

**Procedure**
Determination of the digits of a natural number in the ternary numeral system

The representation of the natural number $a$ in the ternary numeral system can be obtained in the following way:

You look for the highest power of 3, which is at most equal to $a$, and subtract it from $a$; one thus obtains a number $b$.

Then you look for the highest power of 3, which is at most equal to $b$, and subtract it from $b$; one thus obtains a number $c$ … etc.

**Example: Determining the Digits of the Number $a = 65$ in the Ternary Numeral System**
- Highest possible power of 3, which is at most equal to $a$: 27. You subtract 27 from 65 and you get $b = 65 - 27 = 38$.

- Highest possible power of 3, which is at most equal to $b$: 27. You subtract 27 from 38 to get $c = 38 - 27 = 11$.
- Highest possible power of 3, which is at most equal to $c$: 9. You subtract 9 from 11 and you get $d = 11 - 9 = 2$.
- Highest possible power of 3, which is at most equal to $d$: 1. You subtract 1 from 2 and get $e = 2 - 1 = 1$.
- Highest possible power of 3, which is at most equal to $e$: 1. You subtract 1 from 1 and get $1 - 1 = 0$.

This results in the following representation of the number 65 as the sum of powers of 3:

$$65 = 0 \cdot 3^4 + 2 \cdot 3^3 + 1 \cdot 3^2 + 0 \cdot 3^1 + 2 \cdot 3^0$$

One therefore has: $65_{(10)} = {}_{(3)}02102$.

For the examples from Sects. 9.1 and 9.2 you find the following representations in the ternary numeral system:

$$10 = 0 \cdot 3^4 + 0 \cdot 3^3 + 1 \cdot 3^2 + 0 \cdot 3^1 + 1 \cdot 3^0 = {}_{(3)}00101 \ (above: \mathbf{00101})$$
$$31 = 0 \cdot 3^4 + 1 \cdot 3^3 + 0 \cdot 3^2 + 1 \cdot 3^1 + 1 \cdot 3^0 = {}_{(3)}01011 \ (above: \mathbf{01011})$$
$$8 = 0 \cdot 3^4 + 0 \cdot 3^3 + 0 \cdot 3^2 + 2 \cdot 3^1 + 2 \cdot 3^0 = {}_{(3)}00022 \ (above: \mathbf{0010(-1)})$$
$$17 = 0 \cdot 3^4 + 0 \cdot 3^3 + 1 \cdot 3^2 + 2 \cdot 3^1 + 2 \cdot 3^0 = {}_{(3)}00122 \ (above: \mathbf{01(-1)0(-1)})$$
$$55 = 0 \cdot 3^4 + 2 \cdot 3^3 + 0 \cdot 3^2 + 0 \cdot 3^1 + 1 \cdot 3^0 = {}_{(3)}02001 \ (above: \mathbf{1(-1)001})$$
$$97 = 1 \cdot 3^4 + 0 \cdot 3^3 + 1 \cdot 3^2 + 2 \cdot 3^1 + 1 \cdot 3^0 = {}_{(3)}10121 \ (above: \mathbf{11(-1)(-1)1})$$

## 9.4    Relationship Between the Two Representations

The digits 0, 1, and 2 are used to represent numbers in the ternary numeral system. This indicates whether the different powers of 3 are needed 0 times, 1 time, or 2 times to represent the number.

When weighing with the beam balance, we presumed that each of the five balance weights are only available *once*. If you actually need two balance weights of one kind to weigh an object, you can make up for it by using the next larger balance weight and compensate for the excess on the right by placing a suitable balance weight on the left.

**Example: Weighing an Object of 65 g**

If you get each of the balance weights twice, then an object weighing 65 g could be represented as follows (weighing in the ternary numeral system)

$$65\,g = \mathbf{0} \cdot 81\,g + \mathbf{2} \cdot 27\,g + \mathbf{1} \cdot 9\,g + \mathbf{0} \cdot 3\,g + \mathbf{2} \cdot 1\,g$$

However, since the **27** g and the **1** g balance weight is only available once,
  2·27 g is replaced by 1·81 g − 1·27 g and
  2·1 g is replaced by 1·3 g − 1·1 g so:

$$65\,g = \mathbf{1} \cdot 81\,g + (-\mathbf{1}) \cdot 27\,g + \mathbf{1} \cdot 9\,g + \mathbf{1} \cdot 3\,g + (-\mathbf{1}) \cdot 1\,g$$

---

**Rule**

**From the representation of a natural number in the ternary numeral system to weighing with the beam balance**

Replace the term $\mathbf{2} \cdot \mathbf{3^x}$ of the representation of a natural number in the ternary numeral system by $\mathbf{1} \cdot \mathbf{3^{x+1}} + (-\mathbf{1}) \cdot \mathbf{3^x}$. ◀

The transformation does not always proceed as smoothly as in the example with the number 65, for example, the number 68 requires several intermediate steps:

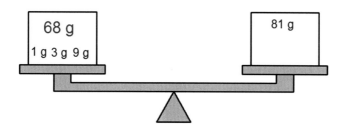

$$68\,g = \mathbf{0} \cdot 81\,g + \mathbf{2} \cdot 27\,g + \mathbf{1} \cdot 9\,g + \mathbf{1} \cdot 3\,g + \mathbf{2} \cdot 1\,g$$
$$= \mathbf{1} \cdot 81\,g + (-\mathbf{1}) \cdot 27\,g + \mathbf{1} \cdot 9\,g + (\mathbf{1} + \mathbf{1}) \cdot 3\,g + (-\mathbf{1}) \cdot 1\,g$$
$$= \mathbf{1} \cdot 81\,g + (-\mathbf{1}) \cdot 27\,g + \mathbf{1} \cdot 9\,g + \mathbf{2} \cdot 3\,g + (-\mathbf{1}) \cdot 1\,g$$
$$= \mathbf{1} \cdot 81\,g + (-\mathbf{1}) \cdot 27\,g + (\mathbf{1} + \mathbf{1}) \cdot 9\,g + (-\mathbf{1}) \cdot 3\,g + (-\mathbf{1}) \cdot 1\,g$$
$$= \mathbf{1} \cdot 81\,g + (-\mathbf{1}) \cdot 27\,g + \mathbf{2} \cdot 9\,g + (-\mathbf{1}) \cdot 3\,g + (-\mathbf{1}) \cdot 1\,g$$
$$= \mathbf{1} \cdot 81\,g + (-\mathbf{1} + \mathbf{1}) \cdot 27\,g + (-\mathbf{1}) \cdot 9\,g + (-\mathbf{1}) \cdot 3\,g + (-\mathbf{1}) \cdot 1\,g$$
$$= \mathbf{1} \cdot 81\,g + \mathbf{0} \cdot 27\,g + (-\mathbf{1}) \cdot 9\,g + (-\mathbf{1}) \cdot 3\,g + (-\mathbf{1}) \cdot 1\,g$$

A fairly large effort is required for the number 41, too:

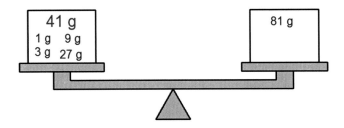

$$41\,g = \mathbf{0} \cdot 81\,g + \mathbf{1} \cdot 27\,g + \mathbf{1} \cdot 9\,g + \mathbf{1} \cdot 3\,g + \mathbf{2} \cdot 1\,g$$
$$= \mathbf{0} \cdot 81\,g + \mathbf{1} \cdot 27\,g + \mathbf{1} \cdot 9\,g + \mathbf{2} \cdot 3\,g + (\mathbf{-1}) \cdot 1\,g$$
$$= \mathbf{0} \cdot 81\,g + \mathbf{1} \cdot 27\,g + \mathbf{2} \cdot 9\,g + (\mathbf{-1}) \cdot 3\,g + (\mathbf{-1}) \cdot 1\,g$$
$$= \mathbf{0} \cdot 81\,g + \mathbf{2} \cdot 27\,g + (\mathbf{-1}) \cdot 9\,g + (\mathbf{-1}) \cdot 3\,g + (\mathbf{-1}) \cdot 1\,g$$
$$= \mathbf{1} \cdot 81\,g + (\mathbf{-1}) \cdot 27\,g + (\mathbf{-1}) \cdot 9\,g + (\mathbf{-1}) \cdot 3\,g + (\mathbf{-1}) \cdot 1\,g$$

*Note:* In the literature, the description of natural numbers using the special balance weights of powers of 3 is also called a **balanced ternary system**.

**Suggestions for Reflection and for Investigations**
**A 9.2**: Determine systematically the object's weights for which one, two, or three balance weights are required for the left scale pan.

## 9.5    References to Further Literature

On **Wikipedia** you can find further information and literature on the keywords in English (German, French):

- Ternary numeral system (Ternärsystem, Système trinaire)

Extensive information can be found on **Wolfram Mathworld** under the keyword:

- Ternary

# Tessellation of Regular 2*n*-Sided Figures with Rhombi

# 10

*Beauty is the first test; there is no permanent place in the world for ugly mathematics.*

*(Godfrey Harold Hardy, British mathematician, 1877–1947).*

Source: https://commons.wikimedia.org/wiki/

In 1973, Roger Penrose discovered a way of covering a plane with *aperiodic* tile patterns so that no patterns got repeated periodically. In the figure above you can see that regular 10-sided figures are included.

In this chapter we will deal – beyond the tessellation of regular 10-sided figures – with a special characteristic of regular 2*n*-sided figures (i.e., polygons with an even number of vertices), namely with the property that they can be completely *tessellated* with rhombi (diamond shapes).

## 10.1   Tessellation of a Regular 10-Sided Figure

As we did in Chap. 1 we start with a mental walk around such a regular figure:

Given a fixed length for the sides, we walk from a starting point to the vertex at the end of the segment, and when we get there, we change our direction by 36°. Then we walk along the next side and repeat this 10 times in total. When we have changed our direction 10 times, we have made a total rotation of 360°, that is, after the 10 turns, we look in the same direction as when we started, and by walking 10 equally long sides we arrive back at the starting point.

From the angle of the change of direction it follows that all interior angles of the regular 10-sided figure have an angular size of 144°.

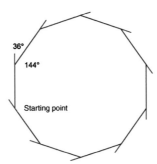

For the following considerations we start at the lower vertex of the 10-sided figure.

A rhombus with the angles 36° (acute angle at the bottom and the top) and 144° (obtuse angle = interior angle of the 10-sided figure) fits on two adjacent sides of the 10-sided figure. In the following we will refer to them as 36°-144° rhombi. And since the interior angles of the 10-sided figure are 144°, four such rhombi fit side by side at the lower vertex.

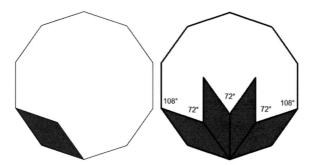

The angle of the gap between two rhombi is $360° - 2 \cdot 144° = 72°$. The angle to the side of the 10-sided figure is $144° - 36° = 108°$ (interior angle of the 10-sided figure = acute angle of the blue-colored rhombus). Then $72°$ and $108°$ are exactly the angles needed for three $72°\text{-}108°$ rhombi, each of which can be inserted.

Above the three $72°\text{-}108°$ rhombi, two $108°\text{-}72°$ rhombi fit, because besides the already placed rhombi an angle of $360° - (2 \cdot 108° + 36°)$ has been left and at the edge there is an angle of $144° - 72° = 72°$, i.e. two rotated $72°\text{-}108°$ rhombi will fit exactly. Finally, there is still a gap for a rotated $36°\text{-}144°$ rhombus (angle at the edge: $144° - 108° = 36°$ and $360° - 3 \cdot 72° = 144°$, respectively).

In total, we need five $36°\text{-}144°$ rhombi and five $72°\text{-}108°$ rhombi for the tessellation of a regular 10-sided figure.

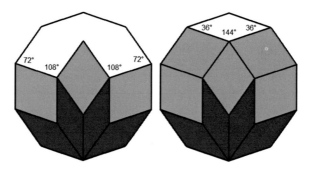

## 10.2 Applying the Method of Tessellation to Other Regular 2n-Sided Figures

In principle this *tree-like* design can be applied to all regular 2n-sided figures.

- $n = 3$: regular 6-sided figure filled with three $60°\text{-}120°$ rhombi,
- $n = 4$: regular 8-sided figure, filled with four $45°\text{-}135°$ rhombi and two $90°\text{-}90°$ rhombi (= squares),
- $n = 5$: regular 10-sided figure, see Sect. 10.1.

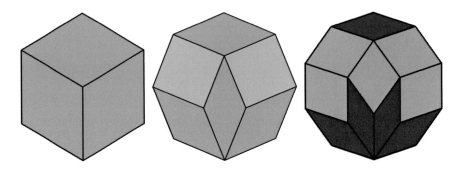

- $n = 6$: regular 12-sided figure, filled with six 30°-150° rhombi, six 60°-120° rhombi, and three 90°-90° rhombi (= squares)
- $n = 7$: regular 14-sided figure, filled with seven 25.71°-154.29° rhombi, seven 51.43°-128.57° rhombi, and seven 77.14°-102.86° rhombi (rounded values)
- $n = 8$: regular 16-sided figure, filled with eight 22.5°-157.5° rhombi, eight 45°-135° rhombi, eight 67.5°-112.5° rhombi, and four 90°-90° rhombi (= squares)

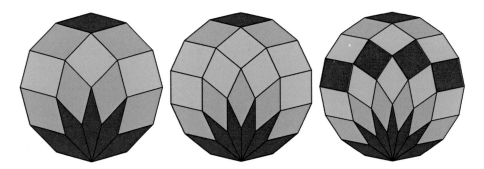

- $n = 9$: regular 18-sided figure, filled with nine 20°-160° rhombi, nine 40°-140° rhombi, nine 60°-120° rhombi, and nine 80°-100° rhombi
- $n = 10$: regular 20-sided figure, filled with ten 18°-162° rhombi, ten 36°-144° rhombi, ten 54°-126° rhombi, ten 72°-108° rhombi, and five 90°-90° rhombi (= squares)

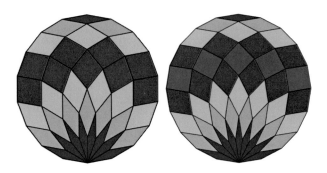

## 10.3 Generalizations of the Tessellation Properties

The investigation of the tree-like tessellations considered so far resulted in the following questions:

- **How many and which rhombi fit into an interior angle of a regular $2n$-sided figure?**

In all $2n$-sided figures considered, the tree-like tessellation starts at the bottom with $n-1$ rhombi, whose acute angle results from the change of direction during the circular walk along the sides of the $2n$-sided figure as $\frac{360°}{2n}$.

The interior angle of the regular $2n$-sided figure is the supplementary angle to the change of direction, that is, $180° - \frac{360°}{2n} = 180° - \frac{180°}{n}$. A rhombus that fits exactly to the edge has, therefore, the angles $180° - \frac{180°}{n}$ and $\frac{180°}{n}$. When you divide the angular size of an interior angle $\left(180° - \frac{180°}{n}\right)$ by the angular size of the acute angle of the narrowest rhombus $\left(\frac{180°}{n}\right)$, then you get exactly $n-1$, that is, exactly $n-1$ of the

$$\left(\frac{180°}{n}\right)\text{-}\left(180° - \frac{180°}{n}\right)\text{rhombi fit into the angle at the bottom.}$$

- **Which types of rhombi are required to fill the gaps?**

In the first layer you start with $n-1$ rhombi whose acute angle has an angular size of $\alpha = \frac{360°}{2n} = \frac{180°}{n}$. It is notable that the rhombi of the following layers have an angle that is double, triple, fourfold, ... as large as the acute angle of the rhombi which lie at the bottom.

The reasons for this can be explained as follows:

- The rhombi of the second layer fit into an angle with
  $360° - 2 \cdot (180° - \alpha) = 2\alpha$ (see the first of the following figures)
- The rhombi of the third layer fit into an angle with
  $360° - 2 \cdot (180° - 2\alpha) - \alpha = 3\alpha$ (see the central one of the following figures)
- The rhombi of the fourth layer fit into an angle with
  $360° - 2(180° - 3\alpha) - 2\alpha = 4\alpha$ (see the last of the following figures)

and so on

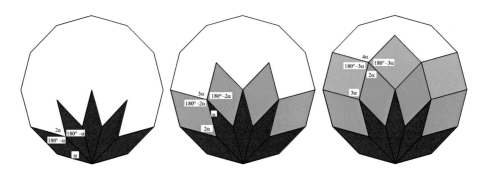

- **How many rhombi are required of each type? How many rhombi are required in total?**

As shown above, in the lowest layer of the $2n$-sided figure we need $(n-1)$ of the $\alpha$ - $(180°{-}\alpha)$ rhombi, where $\alpha = \frac{180°}{n}$. The complementary angle of this rhombus, that is, the angle $180° - \alpha$, can also be displayed as:

$$180° - \alpha = 180° - \tfrac{180°}{n} = 180° \cdot (1 - \tfrac{1}{n}) = \tfrac{180°}{n} \cdot (n-1) = (n-1) \cdot \alpha$$

The same applies accordingly for the other rhombi: $180° - 2\alpha = (n-2) \cdot \alpha$, $180° - 3\alpha = (n-3) \cdot \alpha$ etc.

This means that for the tessellation of the regular $2n$-sided figure rhombi of the same type are required at the "top" and at the "bottom". As there are a total of $n-1$ layers, the "central" layer depends on whether $n$ is even or odd.

In detail:

| Layer No. | Number of rhombi | Angle 1 | Angle 2 |
|:---:|:---:|:---:|:---:|
| 1 | $n-1$ | $1 \cdot \alpha$ | $(n-1) \cdot \alpha$ |
| 2 | $n-2$ | $2 \cdot \alpha$ | $(n-2) \cdot \alpha$ |
| 3 | $n-3$ | $3 \cdot \alpha$ | $(n-3) \cdot \alpha$ |
| ... | ... | ... | ... |
| $n-2$ | 2 | $(n-2) \cdot \alpha$ | $2 \cdot \alpha$ |
| $n-1$ | 1 | $(n-1) \cdot \alpha$ | $1 \cdot \alpha$ |

In total, you lay out $(n-1) + (n-2) + \cdots + 3 + 2 + 1$ rhombi. This number can also be represented using the sum formula $\frac{1}{2} \cdot (n-1) \cdot n$ (see Formula 2.1).

The number of different types of rhombi required grows with a simple regularity:

If $n$ is *odd* the same number of rhombi of each type is required, if $n$ is *even* only half as many rhombi of the last type are required compared to the other types (see the following table).

| $n$ | 3 | 4 | 5 | 6 | 7 | ... |
|---|---|---|---|---|---|---|
| Regular figure | 6-sided | 8-sided | 10-sided | 12-sided | 14-sided | ... |
| $\alpha$ | 60° | 45° | 36° | 30° | 25.71° | ... |
| Number of layers | 2 | 3 | 4 | 5 | 6 | ... |
| Rhombi of Type $\alpha$ | 2 + 1 = 3 | 3 + 1 = 4 | 4 + 1 = 5 | 5 + 1 = 6 | 6 + 1 = 7 | ... |
| Rhombi of Type $2\alpha$ | | 2 | 3 + 2 = 5 | 4 + 2 = 6 | 5 + 2 = 7 | ... |
| Rhombi of Type $3\alpha$ | | | | 3 | 4 + 3 = 7 | ... |
| ... | ... | ... | ... | ... | ... | ... |
| Total number of rhombi | 3 | 6 | 10 | 15 | 21 | ... |

---

**Rule**

**Tree-like tessellation of regular $2n$-sided figures**

Regular $2n$-sided figures can be tessellated with rhombi in a tree-like shape; for this you need a total of $\frac{1}{2} \cdot (n - 1) \cdot n$ rhombi, namely.

- for odd $n$ you need $n$ rhombi with the acute angles $\alpha = \frac{180°}{n}$, $2\alpha$, $3\alpha$, $\ldots$, $\frac{1}{2} \cdot (n - 1) \cdot \alpha$,,
- for even $n$ you need $n$ rhombi with the acute angles $\alpha, 2\alpha, 3\alpha, \ldots, \left(\frac{1}{2} \cdot n - 1\right) \cdot \alpha$ and $\frac{1}{2} \cdot n$ rhombi with the acute angle $\frac{1}{2} \cdot n \cdot \alpha$. ◀

---

## 10.4    Instructions for Making the Diamond Puzzles

Because of the properties described in Sect. 10.3, it is easy to produce the required rhombi. On cardboard (or a similar suitable material) you draw

- The regular $2n$-sided figure,
- $n - 1$ circular arcs with the lower vertex as center, the arcs pass through the other vertices of the $2n$-sided figure, and
- $2n - 3$ line segments that connect the lower vertex with the other vertices of the $2n$-sided figure.

You can draw the rhombi by connecting the points of intersection, see the following figures of the regular 10-sided figure and the 12-sided figure.

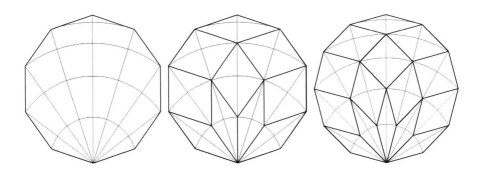

**Suggestions for Reflection and for Investigations**
**A 10.2:** Draw the figure which is needed for the tessellation of a regular 8-sided figure (14-sided figure).

## 10.5    Alternative Tessellation Designs of the Regular 10-Sided Figure with Rhombi

The tree-like design is not the only way to tessellate a regular 2*n*-sided figure.

For example, you can rotate the three non-colored rhombi in the 10-sided figure by 180° as in the following picture, since these three diamonds together form a symmetrical figure. If you perform the rotation, you get an alternative tessellation.

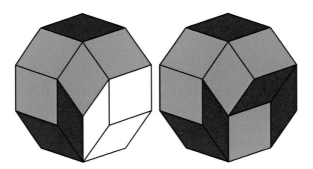

There are a total of six different ways of tessellating the regular 10-sided figure with the 5 + 5 rhombi (see Fig. 10.1a–f).

(*Hint:* We do not differentiate between type 6 and the reflected form).

At the beginning of this section, an example was given of how to change from tessellation type 1 to type 6 by relocating three rhombi – of course, the process can also be reversed. The other types of tessellation can also be obtained by rotating partial areas of the figures by 180°.

The relationships between the different types can be seen in the following diagram.

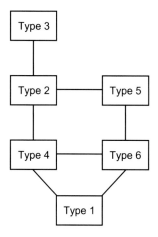

**A 10.3:** What is the relationship between the types of tessellation obtained from rotating the non-colored area? (see the following figures).

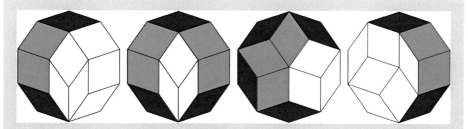

**A 10.4:** Explain: In contrast to the regular 10-sided figure, there is only one way of tessellating the regular 6-sided figure and the regular 8-sided figure.

**A 10.5:** With the regular 12-sided figure, there are plenty of ways of tessellating with the 6+6+3 rhombi. Explore the relationships between the different tessellations (see the following sequence of figures).

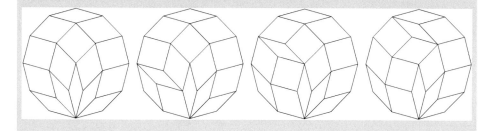

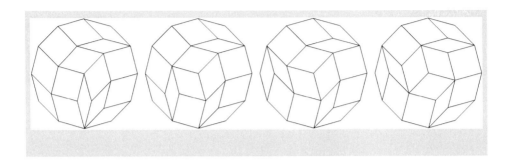

## 10.6  Symmetrical Tessellation of Regular 2n-Sided Figures

Perhaps it is more obvious to start laying out a regular polygon from the center instead of the edge. With a regular 10-sided figure, for example, you start in the center with ten rhombi with an acute angle of $\frac{360°}{10} = 36°$ followed by ten 72°-108° rhombi, ten 108°-72° rhombi, and finally ten 144°-36° rhombi. This also results in a tessellation of a regular 10-sided figure; this, however, consists of four times as many rhombi as the tessellations of 10-sided figures considered so far.

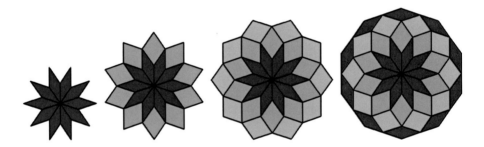

If one starts in the center with five 72°-108° rhombi, then one can again get by with ten rhombi altogether again; it results in type 3 of the regular 10-sided figure (see Fig. 10.1c).

However, if six rhombi with an acute angle $\frac{360°}{6} = 60°$ are laid out in the center, then no regular 12-sided figure, but a regular 6-sided figure results. If you start in the center with seven $\frac{360°}{7}$ rhombi, then you get a regular 14-sided figure. If you place eight $\frac{360°}{8}$ rhombi in the center, then there is no regular 16-sided figure, but a regular 8-sided figure (see the following figures).

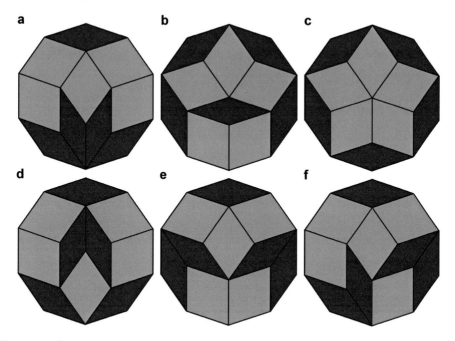

**Fig. 10.1   a–f** The six ways of tessellating a regular 10-sided figure (type 1 to type 6)

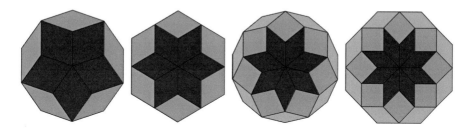

These first experiments with a centrally symmetrical tessellation lead to the following assumption:

Only if *n* is odd, you can lay out the regular 2*n*-sided figure with $\frac{1}{2} \cdot (n - 1) \cdot n$ rhombi in a centrally symmetrical form. For even *n* this will not succeed.

---

**Suggestions for Reflection and for Investigations**
**A 10.6:** Check this assumption also for $n = 9$ and $n = 10$.

---

A hint for this can be already found in the tables in Sect. 10.3: If *n* is odd, then you need the same number of all diamond sizes, with even *n* there is one type of diamonds, of which you only need half as many as the others.

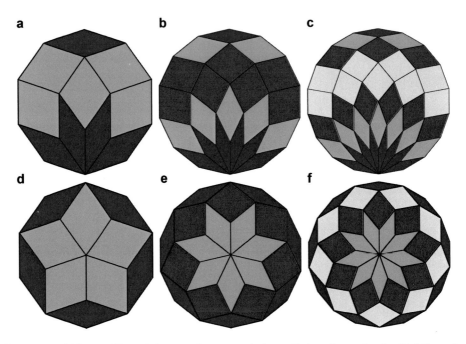

**Fig. 10.2 a–f** The tree like and the centrally symmetrical tessellation of a regular 2*n*-sided figure for *n* = 5, 7, 9

---

**Rule**

**Special feature of the regular 2*n*-sided figures with odd *n***

With regular 2*n*-sided figures with odd *n* both, a tree like and a centrally symmetrical tessellation, is possible. ◀

---

In Fig. 10.2 this special property of the regular 2*n*-sided figure for $n = 5$, $n = 7$, and $n = 9$ is shown.

---

## 10.7   Symmetrical Tessellation of the Regular 2*n*-Sided Figure From Outside to Inside

The fact that a centrally symmetrical design for the 2*n*-sided figures with even *n* does not succeed, is easier to understand when the tessellation begins *at the edge of the figure*. Additionally, you discover a method of tessellating beautiful **star figures**.

- $n = 4$ (regular 8-sided figure), so $\alpha = \frac{180°}{4} = 45°$: you can place the four $\alpha$-$(180° - \alpha)$ rhombi at the edge; but then there is no space for the two $90° - 90°$ rhombi which belong to the tessellation of the regular 8-sided figure.

- $n = 6$ (regular 12-sided figure), so $\alpha = \frac{180°}{6} = 30°$: you can set the six $\alpha$-$(180° - \alpha)$ rhombi at the edge. Then six 90°–90° rhombi fit into the resulting form, but if you take only three rhombi (as in tree-like tessellation), there is no space for the six 60°–120° rhombi.

- $n = 8$ (regular 16-sided figure), so $\alpha = \frac{180°}{8} = 22.5°$: you start with eight $\alpha$-$(180° - \alpha)$ rhombi at the edge, then eight 67.5°–112.5° rhombi fit. For the resulting angles another eight of this kind would be needed, which are not available.

Generalization: As shown in Sect. 10.3, for the tessellation with tree-like patterns, rhombi with the following angles are required: $\alpha$ and 180°–$\alpha$, 2$\alpha$ and 180°–2$\alpha$, 3$\alpha$ and 180°–3$\alpha$, and so on.

The interior angles at the edge of a regular 2$n$-sided figure have an angular size of $180° - \frac{180°}{n} = 180° - \alpha$.

If you put the $n$ rhombi with the acute angle $\alpha = \frac{180°}{n}$ all around the edge, then angles of $(180° - \alpha) - 2\alpha = 180° - 3\alpha$ appear between each of two adjacent rhombi.

In the third layer a rhombus must then fit into the angle $360° - [2 \cdot 3\alpha + (180° - \alpha)] = 180° - 5\alpha$, in the fourth layer into the angle $360° - [2 \cdot 5\alpha + (180° - 3\alpha)] = 180° - 7\alpha$ and so on.

Therefore, for a centrally symmetrical design, one needs rhombi with the angles $\alpha$, $3\alpha$, $5\alpha$, $7\alpha$, and so on.

If both a tree-like design and a centrally symmetrical tessellation of a regular $2n$-sided figure should be possible, then rhombi with the acute angles $\alpha$, $2\alpha$, $3\alpha$, $4\alpha$, …, that is, *all* multiples of $\alpha$ have to occur.

For $n = 5$ (two layers) is $\alpha = 36^\circ$ and you start at the edge with the $\alpha$-$4\alpha$-rhombi, then in the second layer the $3\alpha$-$2\alpha$-rhombi, that means, all angles $\alpha$, $2\alpha$, $3\alpha$, and $4\alpha$ occur.

For $n = 7$ (three layers) you start at the edge with the $\alpha$-$6\alpha$-rhombi, followed by the $3\alpha$-$4\alpha$-rhombi and finally the $5\alpha$-$2\alpha$-rhombi, that means, all angles $\alpha$, $2\alpha$, $3\alpha$, $4\alpha$, $5\alpha$, and $6\alpha$ occur.

For $n = 9$ (four layers) the $\alpha$-$8\alpha$-rhombi, $3\alpha$-$6\alpha$-rhombi, $5\alpha$-$4\alpha$-rhombi, and $7\alpha$-$2\alpha$-rhombi are laid out one after the other from the edge to the center, that means, again all angles $\alpha$, $2\alpha$, $3\alpha$, $4\alpha$, $5\alpha$, $6\alpha$, $7\alpha$, and $8\alpha$ occur.

In contrast, for $n = 4$ (two layers) after laying the first layer one does not get further with $\alpha$-$3\alpha$-rhombi,

for $n = 6$ (three layers), after laying out the $\alpha$-$5\alpha$-rhombi and $3\alpha$-$3\alpha$-rhombi,

for $n = 8$ (four layers), after laying out the $\alpha$-$7\alpha$-rhombi and the $3\alpha$-$5\alpha$-rhombi one no longer has matching rhombi.

Nevertheless, Fig. 10.3a–c shows that symmetrical tessellations also exist for regular 8-sided figures, 12-sided figures, 16-sided figures, etc.

A regular 8-sided figure, 12-sided figure, 16-sided figure, etc. with symmetrical design is only possible with four times the number of rhombi, that is, with $4 \cdot \frac{1}{2} \cdot (n - 1) \cdot n$ rhombi, so for

- $n = 4$ (regular 8-sided figure) with $16 + 8 = 24$ rhombi,
- $n = 6$ (regular 12-sided figure) with $24 + 24 + 12 = 60$ rhombi,
- $n = 8$ (regular 16-sided figure) with $32 + 32 + 32 + 16 = 112$ rhombi.

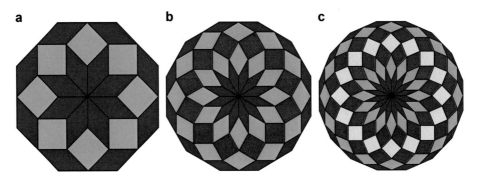

**Fig. 10.3  a–c** Centrally symmetrical $2n$-sided figures with four times the number of rhombi ($n = 4$, $n = 6$, $n = 8$)

## 10.8   Rhombus Tessellation for Regular 5-Sided Figures, 7-Sided Figures, 9-Sided Figures, etc.

An argument to show that there is no tessellation with rhombi for the regular polygons with an *odd* number of vertices can be deduced directly from the considerations of the last section: a centrally symmetrical tessellation from the edge to the center is not possible, because a layout of rhombi at the edge of a regular $n$-sided figure requires that $n$ is an even number, since each rhombus is touching on two sides of the $n$-sided figure. In addition, after the first step it is already apparent that this cannot be continued, see the following figures of the regular 5-sided figure, 7-sided figure, and 9-sided figure.

Another amazing approach results if your rotate the regular $n$-sided figures:

If you draw a regular polygon with $n$ vertices (drawn in red in the following pictures) and rotate it $n$-times by $\frac{360°}{n}$, then the regular $n$-sided figures for *even n* form the centrally symmetrical figures considered in Sect. 10.5.

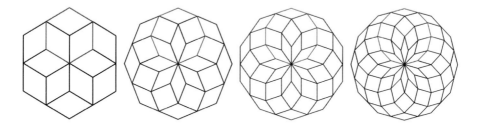

The red framed areas contain exactly the tree-like tessellation, which we have seen in Sect. 10.2. This is shown in the following figures (for $n = 6, 8, 10, 12$) even more clearly.

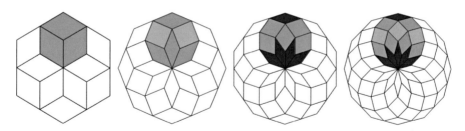

In contrast, for *odd n* no structure is recognizable (see the following figures for $n = 5, 7, 9, 11$).

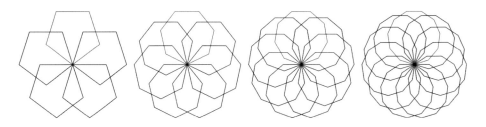

**Suggestions for Reflection and for Investigations**
**A 10.7:** Analyze the following centrally symmetrical star figures tessellated with rhombi.

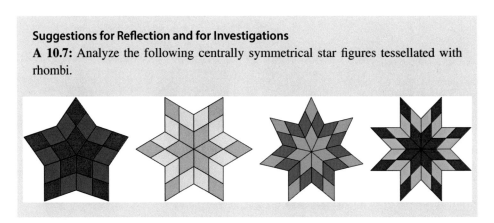

## 10.9    References to further literature

On **Wikipedia** you will find further information and literature on the keyword in English (German, French):

- Penrose Tiling* (Penrose-Parkettierung, Pavage de Penrose)
  *) *Marked as excellent article*

In the Wikipedia article *Euclidean tilings by convex regular polygons* you find an overview of the tessellation of the plane with regular convex polygons.

The following book also contains some explanations of tessellation with rhombi:
- Walser, Hans (2014): *Symmetrie in Raum und Zeit.* Leipzig, Edition am Gutenbergplatz

# Geometric Figures on Grid Paper

**11**

*You have to guess a mathematical theorem before you prove it, you have to guess the idea of a proof before you perform the details.*

*(George Pólya, American mathematician of Hungarian origin, 1887–1985).*

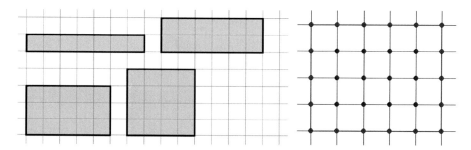

The illustration on the left shows rectangles that can be drawn on paper ruled in squares without any special tools. Instead of this special paper, you can generally use a square grid to draw simple geometric figures, for example, you use a coordinate system with integral coordinates.

In this chapter, we will first deal with rectangles on paper ruled in squares, which we will then modify step-by-step. Then we are going to explore an easy way of calculating the area of figures whose vertices lie on grid points.

© Springer-Verlag GmbH Germany, part of Springer Nature 2021
H. K. Strick, *Mathematics is Beautiful*, https://doi.org/10.1007/978-3-662-62689-4_11

## 11.1   Rectangles with a Given Area

To create a square (a rectangle with sides of equal length) with the area $A = 1$ (area unit), 4 lines are drawn, each of length 1. The length of the line that surrounds the geometrical object is the **perimeter of the figure.**

When drawing rectangles on a paper ruled in squares, given an area $A$ of 1, 2, or 3 area units, the rectangles (except for rotation by 90°) are clearly defined:

The rectangle with the area

$A = 1$ has the perimeter $p = 1+1+1+1 = 4$,
$A = 2$ has the perimeter $p = 2+1+2+1 = 2 \cdot (2+1) = 6$,
$A = 3$ has the perimeter $p = 3+1+3+1 = 2 \cdot (3+1) = 8$.

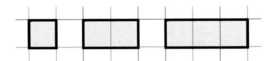

There are two different rectangles with the area $A = 4$ (we call them "equally sized"), namely the rectangle

on the left with the perimeter $p = 4+1+4+1 = 2 \cdot (4+1) = 10$,
and on the right with the perimeter $p = 2+2+2+2 = 4 \cdot 2 = 8$.

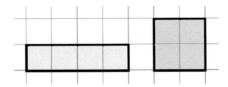

The same applies to rectangles with the area $A = 6$. Again, there are two different rectangles, namely the rectangle

on the left with the perimeter $p = 6+1+6+1 = 2 \cdot (6+1) = 14$,
on the right with the perimeter $p = 3+2+3+2 = 2 \cdot (3+2) = 10$.

**Suggestions for Reflection and for Investigations**
**A 11.1:** Show that there are also two different rectangles for the area $A = 8$, $A = 9$, and $A = 10$. Draw these figures.

In the cases $A = 4$, $A = 6$, $A = 8$, and $A = 10$ that two rectangles exist is related to the fact that the natural numbers 4, 6, 8, and 10 can be divided by 2, so that in addition to the rectangles with the formats $4 \times 1$, $6 \times 1$, $8 \times 1$, and $10 \times 1$, there are also rectangles with the formats $2 \times 2$, $3 \times 2$, $4 \times 2$, and $5 \times 2$. And since the natural number 9 can be divided by 3, there is also a rectangle in the format $3 \times 3$ in addition to the rectangle in the format $9 \times 1$.

Instead of the typical spelling for rectangles, with the "$\times$" symbol, there is also the way of writing these cases as **pairs of divisors,** for example, the natural number 4 can be related to the pairs $(1, \ 4)$ and $(2, \ 2)$.

**A 11.2:** Which are the pairs of divisors to the areas 6, 8, 9, 10? Provide the corresponding rectangular perimeters.

The natural number 12 has three pairs of divisors, namely $(1, 12)$, $(2, 6)$, and $(3, 4)$. Therefore, there are three different rectangles with the area $A = 12$. The perimeters of these three rectangles are $p = 2 \cdot (12 + 1) = 26$, $p = 2 \cdot (6 + 2) = 16$, and $p = 2 \cdot (4 + 3) = 14$

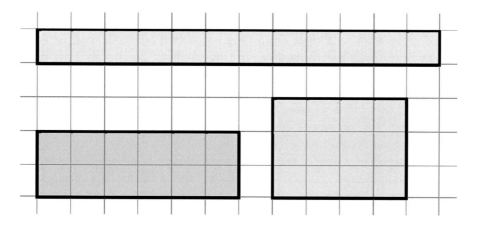

From the previous considerations a rule becomes clear, which is illustrated by Fig. 11.1.

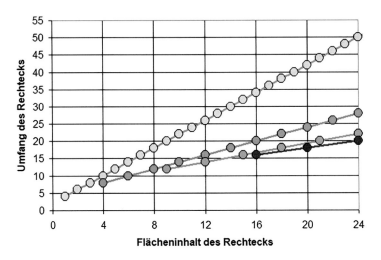

**Fig. 11.1**  Relationship between area and perimeter of rectangles

---

**Rule**

**Number of rectangles with a given integral area**

To check how many rectangles with a given area $A$ exist, you have to determine all pairs of divisors of the natural number $A$.

- To every natural number $n$ with $n \geq 1$ there is a pair of divisors $(1;\ n)$, which is related to a rectangle with the side lengths 1 and $n$, thus with the area $A = n \cdot 1 = n$ and the perimeter $p = 2 \cdot (n+1) = 2n+2$
- To any even natural number $n = 2m$ with $n \geq 4$ there is an additional pair of divisors $(2;\ m)$, and this is related to a rectangle with the side lengths 2 and $m$, thus with the area $A = m \cdot 2 = 2m = n$ and the perimeter $p = 2 \cdot (m+2) = 2m+4$
- For every natural number that is divisible by 3, i.e. $n = 3m$ with $n \geq 9$, there is an additional pair of divisors $(3;\ m)$, and this is related to a rectangle with side lengths 3 and $m$, thus with the area $A = m \cdot 3 = 3m = n$ and the perimeter $p = 2 \cdot (m+3) = 2m+6 = \frac{2}{3}n + 6$.
- For every natural number that is divisible by 4, i.e. $n = 4m$ with $n \geq 16$, there is an additional pair of divisors $(4;\ m)$, and this is related to a rectangle with side lengths 4 and $m$, thus with the area $A = m \cdot 4 = 4m = n$ and the perimeter $p = 2 \cdot (m+4) = 2m+8 = \frac{2}{4}n + 8 = \frac{1}{2}n + 8$. ◀

---

**Suggestions for Reflection and for Investigations**

**A 11.3:** Why are restrictions needed in the rule ($n \geq 1, n \geq 4, n \geq 9$ etc.)?

## 11.2   Rectangles of Equal Perimeter

When looking at the rectangles considered so far, it is noticeable that there are two rectangles that have the perimeter $p=8$, namely the $3 \times 1$ rectangle (with an area of 3) and the $2 \times 2$ rectangle (with an area of 4).

And for the perimeter $p=10$, you will find the $4 \times 1$ rectangle (with an area of 4) and the $3 \times 2$ rectangle (with an area of 6).

For the perimeter $p=12$ there are three rectangles, namely the $5 \times 1$ rectangle (with an area of 5), the $4 \times 2$ rectangle (with an area of 8), and the $3 \times 3$ rectangle (with an area of 9).

We will now explore in general how many different rectangles exist for a given perimeter.

> **Example: Rectangles with the Perimeter p=20**
> The following graphics show the five rectangles of different size with perimeter $p=20$, that is, with the side lengths $a$ and $b$ with $a + b = 10$. The $9 \times 1$ rectangle (with the area 9), the $8 \times 2$ rectangle (with the area 16), the $7 \times 3$ rectangle (with the area 21), the $6 \times 4$ rectangle (with the area 24), and the $5 \times 5$ rectangle (with the area 25).

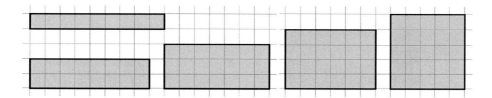

Generally, all rectangles on paper ruled in squares have a perimeter that is even, that is, divisible by 2. This is due to the fact that two opposite sides of each rectangle are equal in size and, therefore, the corresponding lengths occur twice in the sum which must be calculated for the perimeter.

Therefore, the simplest way to proceed is as follows:

- Halve the natural number $p$ and determine all pairs of natural numbers $(a, b)$ with $a \leq b$ for which the following condition applies: $a + b = \frac{1}{2} \cdot p$.

**Examples with $p=16$ and for $p=18$**

$p=16$, so $\frac{1}{2} \cdot p = 8$: pairs of numbers: $(1,7)$, $(2,6)$, $(3,5)$, $(4,4)$ see the following graphics on the left.

$p=18$, so $\frac{1}{2} \cdot p = 9$: pairs of numbers: $(1,8)$, $(2,7)$, $(3,6)$, $(4,5)$ see the following graphics on the right.

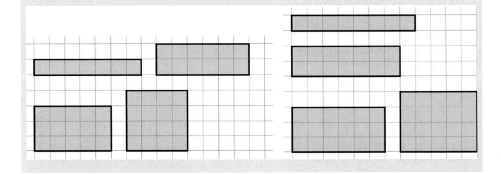

To find a rule for how many pairs of numbers exist for a certain perimeter $p$ we consider the following table in Fig. 11.2.

If you compare the yellow highlighted rows of the table with the green highlighted rows, the following rule becomes clear.

| Perimeter $p$ | Pairs of numbers | Number of possible rectangles |
|:---:|:---|:---:|
| 4 | (1 , 1) | 1 |
| 6 | (1 , 2) | 1 |
| 8 | (1 , 3), (2 , 2) | 2 |
| 10 | (1 , 4), (2 , 3) | 2 |
| 12 | (1 , 5), (2 , 4), (3 , 3) | 3 |
| 14 | (1 , 6), (2 , 5), (3 , 4) | 3 |
| 16 | (1 , 7), (2 , 6), (3 , 5), (4 , 4) | 4 |
| 18 | (1 , 8), (2 , 7), (3 , 6), (4 , 5) | 4 |
| 20 | (1 , 9), (2 , 8), (3 , 7), (4 , 6), (5 , 5) | 5 |
| 22 | (1 , 10), (2 , 9), (3 , 8), (4 , 7), (5 , 6) | 5 |
| ... | ... | ... |

**Fig. 11.2** Number of rectangles with a given integral perimeter

---

**Rule**

**Number of rectangles with a given integral perimeter**

- If the perimeter $p$ is a natural number divisible by 4, then a quarter of $p$ is exactly equal to the number of possible rectangles with this perimeter.
- The number of possible rectangles remains the same, if the perimeter of the rectangle is larger by 2.

Furthermore:

- If the perimeter $p$ is a natural number divisible by 4, then one of the *possible* rectangles is a square; in this case a quarter of the perimeter is equal to the side length of the square. ◀

Finally, we explore the **maximum** area that can be covered by a rectangle whose perimeter is given.

The following graphics for $p=16$ (yellow), $p=18$ (green), $p=20$ (orange–red), and $p=22$ (blue) are obtained by determining the area of each rectangle in the table from Fig. 11.2.

In contrast to the previous considerations, the rectangles with the side lengths $a$ and $b$, which are obtained by *swapping*, are also plotted.

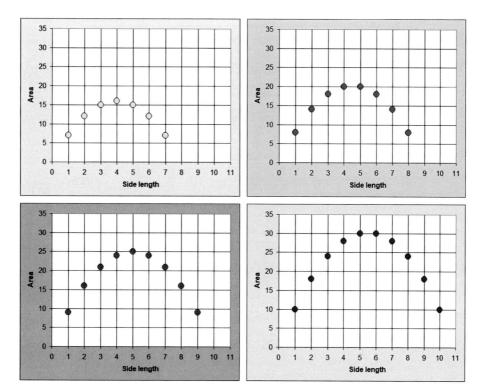

The points lie on **quadratic parabolas** opened up downwards.

If the perimeter $p$ is divisible by four, the maximum (i.e., the largest possible area) exists and all four sides of the rectangle are of equal length, that is, you get a square.

(By the way, this also applies generally to rectangles that do not have integral side lengths).

---

**Theorem**

**Rectangle with maximum area**

Among all rectangles with a given perimeter, the square has the largest area.

Conversely, the square has the smallest perimeter of all rectangles with a given area. ◀

---

**Suggestions for Reflection and for Investigations**

**A 11.4:** Explain Fig. 11.3.

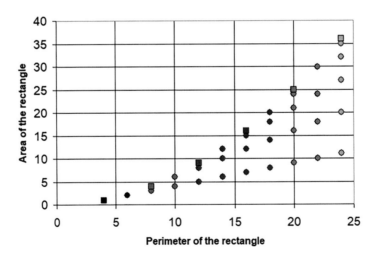

**Fig. 11.3** Areas of rectangles with a given perimeter

## 11.3   Special Rectangles: The 4 × 4-Rectangle and the 3 × 6-Rectangle

Perhaps when looking at the rectangles with perimeter 16 and 18, you may have noticed that there is a rectangle for which the numerical values for $A$ and $p$ match:

$A = 16$ (area units) and $p = 16$ (length units) or $A = 18$ (area units) and $p = 18$ (length units).

These are the 4 × 4-rectangle and the 6 × 3-rectangle.

One can prove that only for the two natural numbers 16 and 18 is there a rectangle for which the perimeter and area have the same numerical values.

In Fig. 11.4 these two points $(A, p) = (16,16)$ and $(A, p) = (18,18)$ are highlighted; the straight line is also drawn, on which the points correspondent to equal numerical values.

(16,16) and (18,18) seem to be the only points with integral coordinates on the straight line.

This is of course not a complete proof, because the graphic shows only a limited range. For larger perimeters there might be other rectangles with this special feature.

That this is indeed not possible can be shown as follows:

For the side lengths $a$, $b$ must apply $2 \cdot (a + b) = a \cdot b$.

This equation can be transformed:

$2 \cdot a = a \cdot b - 2 \cdot b \Leftrightarrow 2 \cdot a = b \cdot (a - 2) \Leftrightarrow b = \frac{2a}{a-2} = \frac{2 \cdot (a-2)+4}{a-2} = 2 + \frac{4}{a-2}$ (for $a > 2$).

You can illustrate this relationship in a $a$–$b$-coordinate system, see the following figure.

You can see that only the three points on the graph (3, 6), (4, 4), and (6, 3) have integral coordinates.

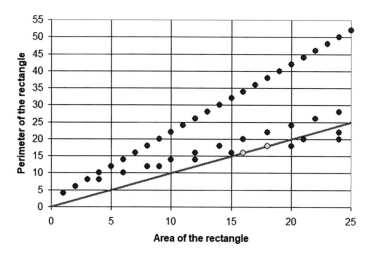

Fig. 11.4   Relationship between the area and perimeter of rectangles

That there are no other points outside the chosen range can be seen if you look at the strict monotonicity of the graph:

For $a > 6$ applies (since the denominator term is in any case greater than zero):

$2 < b = 2 + \frac{4}{a-2} < 3$, that is, all values of the function lie between 2 and 3,

So outside of the range shown in the graphics, there are no more integral function values.

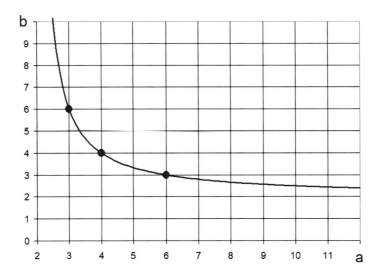

## 11.4   Variations of Rectangular Figures

If you remove one unit square or several unit squares next to a corner of a rectangle, the area is reduced by a corresponding number of area units – but the perimeter remains the same.

**Example**
From the $3 \times 2$-rectangle with an area of 6 area units, you can at most remove two unit squares at the corners (see the following figures with red dotted lines), that is, for the area $A$ of these figures with a perimeter of 10 we have $4 \leq A \leq 6$.

Left figure: $A = 2 \cdot 3 - 1 = 5$; $p = 2 \cdot (2+3) = 10$,
Other figures: $A = 2 \cdot 3 - 2 = 4$; $p = 2 \cdot (2+3) = 10$.

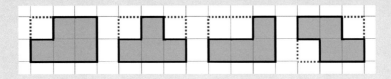

From the $3 \times 3$-rectangle with an area of 9, you can remove at most four unit squares at the corners, that is, for the area $A$ of these figures with a perimeter of 12 we have $5 \le A \le 9$.

For all these figures you get $A = 3 \cdot 3 - 4 = 5; p = 2 \cdot (3+3) = 12$.

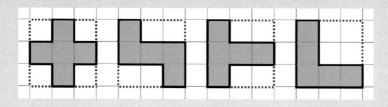

**Suggestions for Reflection and for Investigations**
**A 11.5:** How many unit squares can at most be removed at the corners of an $a \times b$ rectangle?

However, if you remove one or more unit squares that do *not* lie at a corner, it gets more complicated....

The following examples show that the perimeter increases by 2, 4, 6, ... if you remove squares from the edge (not at the corners). In the graphics, the added lines are shown in violet.

You can also see from the following examples that the number of unit squares that are removed plays an important role, but the "depth" of the "cut-out" is essential.

**Examples of Changes by Removing Squares**
For the following rectangles we have (from left to right)

$A = 2 \cdot 3 - 1 = 5; p = 2 \cdot (2+3) + 2 = 12,$
$A = 3 \cdot 3 - 2 = 7; p = 2 \cdot (3+3) + 4 = 16,$
$A = 3 \cdot 3 - 2 = 7; p = 2 \cdot (3+3) + 4 = 16$
$A = 2 \cdot 4 - 2 = 6; p = 2 \cdot (2+4) + 2 = 14.$

For     each     of     the     following     rectangles     we     have:
$A = 3 \cdot 4 - 3 = 9; \; p = 2 \cdot (3 + 4) + 6 = 20.$

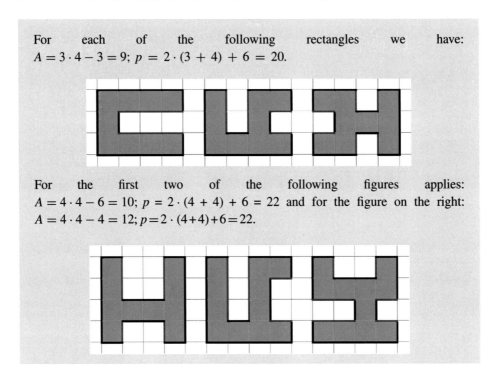

For     the     first     two     of     the     following     figures     applies:
$A = 4 \cdot 4 - 6 = 10; \; p = 2 \cdot (4 + 4) + 6 = 22$ and for the figure on the right:
$A = 4 \cdot 4 - 4 = 12; \; p = 2 \cdot (4+4)+6 = 22.$

**Suggestions for Reflection and for Investigations**

**A 11.6:** By how much can the perimeter of an $a \times b$ rectangle be increased *at most* by removing unit squares?

**A 11.7:** To what extent can areas and perimeters ($=$ length of the boundary lines inside *and* outside) be changed by removing unit squares from the *inside* of the rectangles, see the following illustrations.

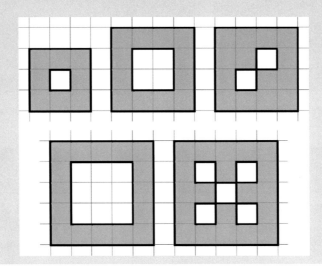

**A 11.8:** Explore which shapes can be created by removing unit squares starting with an initial figure, for example, a $3 \times 3$-rectangle (square). Determine the perimeter and the area of each figure.

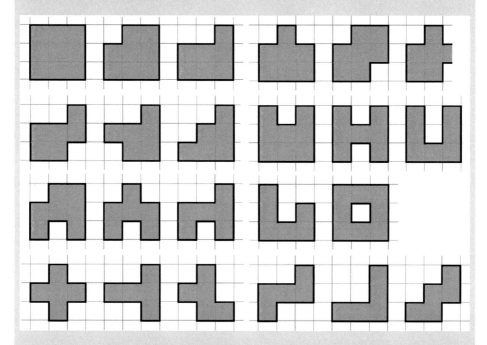

*Hint:* The last shown figures are **pentominoes,** see Chap. 5.

Further figures can be obtained from the figures shown by rotation or by reflection.

**A 11.9:** Which figures result from a $4 \times 3$-rectangle as initial figure? Determine the perimeter and area of each possible figure.

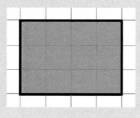

### Rectangles with an Integral area but not an Integral Perimeter

To draw the following squares we use the grid points of grid paper, but not the lines. Although the vertices of the squares in the following graphics are rotated compared to the grid lines, we can still calculate the areas:

You only need to draw a frame around the squares, that is, a square with the vertices of the given square on its sides, see the following figures.

**Suggestions for Reflection and for Investigations**

**A 11.10:** What is the area of the light blue squares? Are there simple rules for calculating these areas? Which area sizes can occur?

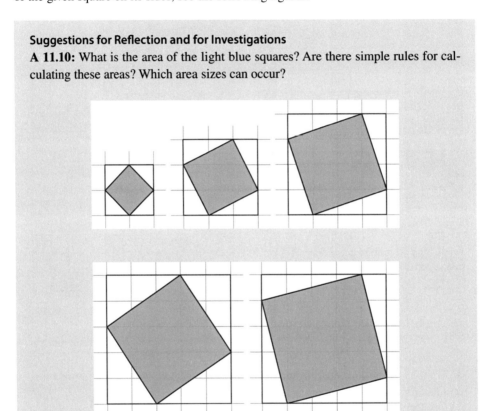

**A 11.11:** In a similar manner as in **A 11.10** you can also draw rotated *rectangles* in such a way that their vertices lie on the grid points. What do you have to attend to here? Which areas can occur?

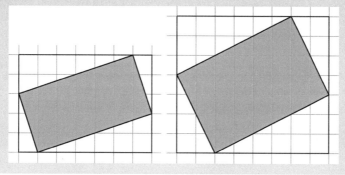

*Hint:* A special case is given in the following example: Not only the area but also the perimeter is integral; as the side lengths of the triangles form a Pythagorean triple, see Sect. 2.7.

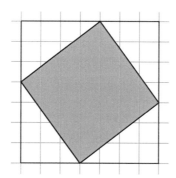

## 11.5  Investigations on Pick's Theorem

Georg Alexander Pick (born 1859 in Vienna) was professor of mathematics at the University of Prague from 1888 to 1927. In 1942 he was killed by the National Socialists in the concentration camp Theresienstadt.

He discovered a simple way to calculate the area of polygons when the vertices of the polygon are points of a grid paper.

---

**Theorem**

**Area of polygons in the square grid**

If the vertices of a polygon lie on a grid of equal-distanced points, then the area $A$ of the polygon only depends on the number $b$ of lattice points on the boundary of the polygon, and on the number $i$ of the lattice points in the interior of the polygon, namely:

$$A = i + \tfrac{1}{2}b - 1 \blacktriangleleft$$

In the literature one can find short and elegant proofs of the formula. Let us approach the statement of the theorem and the proof step by step. An elegant path is generally characterized by the fact that it is straight and avoids detours – which is in contrast to the path taken here, where experience is to be gained by varying tasks. It is accepted that considerations later turn out to be superfluous, but this is also part of the chosen method.

We start with a study of **rectangles.**

Both for calculating the area ($A =$ width $a$ of the rectangle × height $h$ of the rectangle) as well as the perimeter ($u = 2 \cdot$ (width $a$ of the rectangle + height $h$ of the rectangle)) you need the sizes $a$ and $h$.

Therefore, we first include these two variables in our considerations.

Let's start with a rectangle without interior points, that is, we examine rectangles of height 1.

A rectangle of the width 1, that is, a square, has $b = 4$ boundary points. If width $a$ grows step by step by 1 length unit, then 2 boundary points are added in each case. The number $b$ of boundary points of a rectangle with height $h = 1$ is, therefore, calculated according to the formula $b = 2a + 2$.

This can also be formulated as follows: The number $b$ of boundary points at the upper and lower boundary is by 1 greater each than the width $a$, so $b = 2\,(a + 1)$.

If we solve this equation for a we have $a = \tfrac{1}{2}b - 1$.

And since rectangles of height $h = 1$ have an area $A = a$, the following applies:

$$A = \tfrac{1}{2}\,b - 1 \tag{11.1}$$

In other words: If you halve the number $b$ of the boundary points and reduce the result by 1, you get the area $A$ of a rectangle with the height $h = 1$ (see also the following table).

| Width $a$ of the rectangles (in length units) | 1 | 2 | 3 | 4 | 5 | ... |
|---|---|---|---|---|---|---|
| Number $r$ of boundary points | | 4 | 6 | 8 | 10 | 12 | ... |
| Area $A$ (in area units) | | 1 | 2 | 3 | 4 | 5 | ... |

Next, we will examine rectangles with interior points.

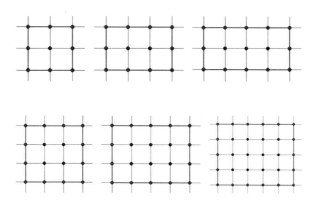

These interior points themselves also form a rectangle. The number $i$ of interior points is obviously smaller by one than the width a and the height $h$ each, so the following is true:

$$i = (a - 1)(h - 1)$$

As the height increases, the number $b$ of the boundary points grows by 2, i.e. you get:

$$b = 2(a + 1) + 2(h - 1) = 2(a + h)$$

When you compare this with the previous rule, you see that the area $A$ of the rectangles differs from the result $A = \frac{1}{2}b - 1$ by the number $i$ of the interior points (see the following table), that is, the following formula applies to rectangles:

$$A = i + \tfrac{1}{2}b - 1 \tag{11.2}$$

| Width a of the rectangles (in length unites) | 2 | 3 | 4 | ... | 3 | 4 | 5 | ... | 4 | 5 | 6 | ... |
|---|---|---|---|---|---|---|---|---|---|---|---|---|
| Height $h$ of the rectangles (in length unites) | 2 | 2 | 2 | ... | 3 | 3 | 3 | ... | 4 | 4 | 4 | ... |
| Number b of boundary points | 8 | 10 | 12 | ... | 12 | 14 | 16 | ... | 16 | 18 | 20 | ... |
| Term to compare ½ · $b$ − 1 | 3 | 4 | 5 | ... | 5 | 6 | 7 | ... | 7 | 8 | 9 | ... |
| Number $i$ of interior points | 1 | 2 | 3 | ... | 4 | 6 | 8 | ... | 9 | 12 | 15 | ... |
| Area A (in area units) | 4 | 6 | 8 | ... | 9 | 12 | 15 | ... | 16 | 20 | 24 | ... |

## 11.6   A Rule for Rectangular Polygons

In the next step, we look at rectangular polygons, where a square has been removed from one of the corners.

By this, an interior point is lost, but the number of vertices remains the same; the area decreases by 1, that is, the formula (11.2) for calculating the area can also be applied without change.

**Suggestions for Reflection and for Investigations**
**A 11.12:** Instead of removing a square at a corner, you can also attach one square at one side (see the following figure). Why does formula (11.2) apply in this case, too?

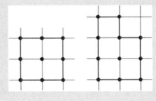

You can remove two or more squares from a corner; the length of the boundary and thus the number of boundary points remains unchanged. With each square removed, one interior point is lost, and the area decreases by 1 each.

**Provisional result:** *If one or more squares are removed at a corner, the number of boundary points remains the same. The area of the figure is reduced by as many units as interior points disappear.*

The situation seems to be different if you do not remove a square at a corner of the rectangle, but at a side of the rectangle, thus reducing the area by 1.

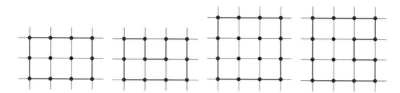

The perimeter of the polygon increases; two of the interior points become boundary points, so we have:

$$A_{\text{new}} = (i - 2) + \tfrac{1}{2} \cdot (b + 2) - 1 = \left(i + \tfrac{1}{2} \cdot b - 1\right) - 1 = A_{\text{previously}} - 1$$

This means that the area of the polygon which is reduced by 1 square unit can also be calculated using the formula (11.2).

**Suggestions for Reflection and for Investigations**

**A 11.13:** Explain: Even if you remove or attach a square at several places (see the following figures), the formula (11.2) is still correct.

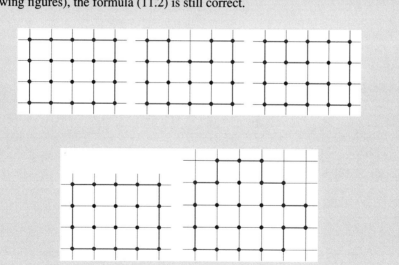

## 11.7    Checking Pick's Theorem for Triangles

In our next step we are going to analyze the situation when we make sloping cuts.

If a right-angled triangle is cut off at a corner of a rectangle, the number of boundary points is reduced. With each omitted boundary point the area is reduced by half an area unit, which is exactly as contained in formula (11.2).

If you cut out a right-angled triangle from a rectangle without removing a corner, then an interior point can become a boundary point, and thus the area is reduced by $\frac{1}{2}$. This also applies if more than one triangle is cut out: With each omitted boundary point the area is reduced by $\frac{1}{2}$.

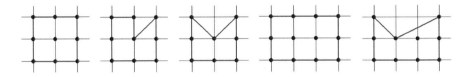

**Suggestions for Reflection and for Investigations**

**A 11.14:** You can also think of sloping boundary lines which have been created by attaching one or more triangles on top of an existing given rectangle (see the following illustrations).

Explain: Formula (11.2) still applies in this case, because the area of the figure increases with each boundary point by $\frac{1}{2}$.

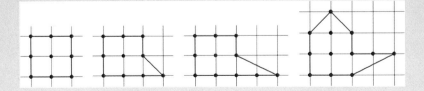

**A 11.15:** The first and the third figures in the following illustrations were created by removing or cutting subareas from the second and the fourth rectangular figures, respectively. Explain the relationship between the areas of the figures.

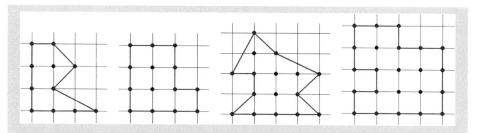

**A 11.16:** Specify rectangular polygons from which the figures in **A 11.15** got created by attaching, and explain the relationship between the areas of the figures.

## 11.8   Considerations on a General Proof of Pick's Theorem

When you consider any polygon whose vertices are grid points, you can subdivide it further and further until only subareas of two basic forms are left: rectangles and the right-angled triangles. In Sect. 11.5 we have explained why Pick's theorem applies to rectangles.

Right-angled triangles in a grid were considered in the previous sections only as attachments to other figures, but not as single figures.

Right-angled triangles with width $a$ and height $h = 1$ have a + 1 grid points below on the boundary and another grid point at the top, so there are $b = a + 2$ boundary points in total, and no interior points.

The area of these triangles is calculated as:

$$A = \tfrac{1}{2}ah = \tfrac{1}{2}a \text{ because h} = 1$$

If one replaces $a = b - 2$ then it results in

$$A = \tfrac{1}{2}(b - 2) = \tfrac{1}{2}b - 1$$

the same formula as for rectangles with the height $h = 1$, see formula (11.1).

Right-angled triangles in which the longest side (hypotenuse) contains a boundary point need not be examined more closely because they can be subdivided into a rectangle and two smaller right-angled triangles (see the following figures).

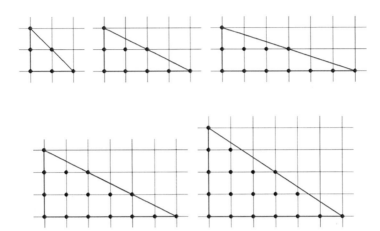

If the width and the height of the right-angled triangles have lengths which are coprime, *none* of the boundary points lies on the hypotenuses. The following illustrations show right-angled triangles with side lengths 2 and 3 or 2 and 5 or 3 and 5, as examples.

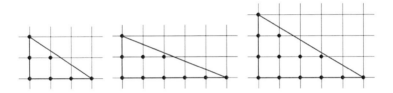

In this case, the triangle has an area that is half the area of the rectangle obtained by reflection at the center of the hypotenuse.

By this mapping the interior points of the right-angled triangle are also reflected (see following example).

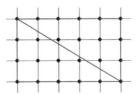

The figure therefore contains twice as many interior points:

$$\tfrac{1}{2} \cdot i_{\text{Rectangle}} = i_{\text{Triangle}}$$

However, the number of boundary points of the rectangle is not twice as large as that of the triangle, but 2 less than the double, because the two end points of the hypotenuse are counted twice when the figure is doubled:

$$b_{\text{Rectangle}} = 2 \cdot b_{\text{Triangle}} - 2$$

We have seen that formula (11.2) is correct for rectangles. It therefore applies:

$$A_{\text{Rectangle}} = i_{\text{Rectangle}} + \tfrac{1}{2} \cdot b_{\text{Triangle}} - 1$$

From this follows for the right-angled triangle, which is half as large as a rectangle:

$$A_{\text{Triangle}} = \tfrac{1}{2} \cdot A_{\text{Rectangle}} = \tfrac{1}{2} \cdot i_{\text{Rectangle}} + \tfrac{1}{4} \cdot b_{\text{Rectangle}} - \tfrac{1}{2}$$

$$= i_{\text{Triangle}} + \left( \tfrac{1}{2} \cdot b_{\text{Triangle}} \right) - \tfrac{1}{2} - \tfrac{1}{2}$$

$$= i_{\text{Triangle}} + \tfrac{1}{2} \cdot b_{\text{Triangle}} - 1$$

The formula (11.2), therefore, also applies to right-angled triangles that have no boundary points on the hypotenuse.

---

**Suggestions for Reflection and for Investigations**
**A 11.17:** Analyze the situation if at least one boundary point lies on the hypotenuse of a right-angled traingle.

---

So at the end, the only thing left to examine is whether the formula is still valid if two rectangles are attached or two right-angled triangles (without a boundary point on the hypotenuse) or a rectangle and a right-angled triangle (without a boundary point on the hypotenuse).

These considerations can be carried out independently, based on examples.

---

**Suggestions for Reflection and for Investigations**
**A 11.18:** Use the following figures to explain why Pick's theorem is valid when two basic figures (as described above) are attached.

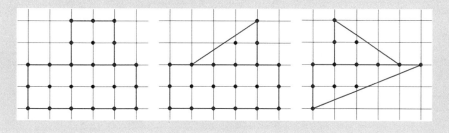

**A 11.19:** Examine whether Pick's theorem can also applied in the following figures.

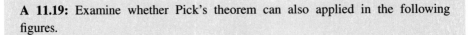

**A 11.20:** As a creative exercise, find symmetrical polygons whose vertices are grid points and which do not have an interior point (which have exactly one interior point). Two examples each are printed below.

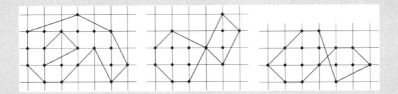

**A 11.21:**

1. The polygons in the following two graphics are determined by points in a *triangular grid*. Calculate the area of these figures. Specify the area in multiples of the area of a grid triangle.

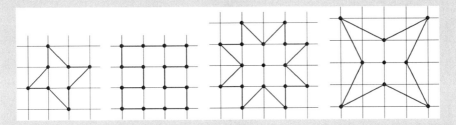

2. Determine for these figures also the number of boundary points and of interior points and check the correctness of the following sentence using these examples:
   - If the vertices of a polygon are grid points of a triangular grid, then the area $A$ of the polygon can be calculated by means of the number $b$ of those boundary points that are grid points and the number $i$ of interior grid points:
     $A = 2i + b - 2$
     when $A$ is measured in multiples of the area of a grid triangle.

3. Explain why the formula $A = 2i + b - 2$ is plausible.

   *Hint*: Consider at first a rhombus grid instead of a square grid (see the following figure on the left) and then the triangular grid, see the following graphics.

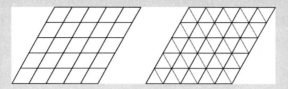

**A 11.22:** Instead of a square grid or triangular grid, you can also use a hexagonal grid to dissect the plane. These grid elements consist of six equilateral triangles each, see the following figure:

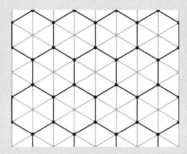

Also with these grids, you can determine the area of a polygon using the number of boundary points and the interior points of the figure. However, counting is a little more complicated – you have to distinguish between two types of boundary points and two types of interior points.

The calculation formula here is as follows:

- If the vertices of a polygon are grid points of a hexagonal grid, then the area $A$ of the polygon can be calculated by means of the number $b_H$ of those boundary points that are grid points and the number $i_H$ of the interior grid points of the hexagonal grid, and the number $b_T$ of the boundary points as well as the number $i_T$ of interior points of the triangular grid that do not belong to the hexagonal grid: $A_H = \frac{1}{3}(i_T + i_H) + \frac{1}{6}(b_T + b_H)$

In doing so, $A$ is measured in multiples of the area of a hexagonal grid unit.

1. In the following figure on the left, a polygon is drawn in the hexagonal grid, whose area obviously equals 3 hexagonal units. Explain the calculation of the area using the formula.

2. The figure on the right has an area of 1.5 hexagonal units. Again, explain how the area is calculated using the formula.
3. How can the modified formula for calculating the area of a polygon in a hexagonal grid be derived?

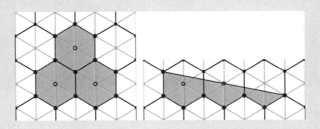

## 11.9   References to Further Literature

On **Wikipedia** you can find further information and literature on the keywords in English (German, French):

- Rectangle (Rechteck, Rectangle)
- Square (Quadrat, Carré)
- Area (Flächeninhalt, Aire)
- Perimeter (Umfang, Périmètre*)
- Pick's theorem (Satz von Pick, Théorème de Pick).
  *) Marked as an excellent article

More informations can be found at **Wolfram Mathworld** under the keyword:

- Rectangle, Square, Area, Perimeter, Pick's Theorem

The interactive website of Alexander Bogomolny offers the chance to draw arbitrary polygons and to check the statement of Pick's theorem:

- https://www.cut-the-knot.org/ctk/Pick.shtml

# Sum of Spots

**12**

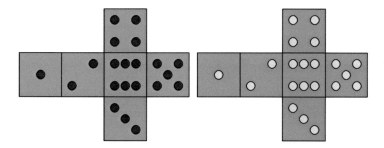

In mathematics, ordinary cubic dice are often called regular hexahedrons – the word is composed of the Greek words *hexa* (Eng. six) and *hedra* (Eng. area).

Usually, the dice are labeled in such a way that the sum of spots of opposite surfaces add up to 7, that is, $7 = 1+6 = 2+5 = 3+4$.

To make a dice, you can use different cube nets. The best known form is the shape of a lying cross used in the figure above, but ten other shapes are also possible (see Chap. 5).

© Springer-Verlag GmbH Germany, part of Springer Nature 2021
H. K. Strick, *Mathematics is Beautiful*, https://doi.org/10.1007/978-3-662-62689-4_12

## 12.1   Sum of Spots When Rolling two Regular Hexahedrons

To determine the possible sums of spots when throwing two ordinary dice labeled in the usual way, make a **table of combinations**. The number of spots of the first dice (e.g., colored green in the figure above) are entered in the left column of the table and the number of spots of the second dice (e.g., blue) in the upper row. Instead of using two different dice, you can also roll one dice twice (see Sect. 12.3).

|   | 1 | 2 | 3 | 4 | 5 | 6 |
|---|---|---|---|---|---|---|
| 1 | 2 | 3 | 4 | 5 | 6 | 7 |
| 2 | 3 | 4 | 5 | 6 | 7 | 8 |
| 3 | 4 | 5 | 6 | 7 | 8 | 9 |
| 4 | 5 | 6 | 7 | 8 | 9 | 10 |
| 5 | 6 | 7 | 8 | 9 | 10 | 11 |
| 6 | 7 | 8 | 9 | 10 | 11 | 12 |

From the combination table of rolling two dice or one dice twice, you can see which of the $6 \cdot 6 = 36$ combinations can result in which sums of spots.

If the dice are fair, that is, if you have no reason to doubt that each number on both dice has the same chance of occuring, then you can also read out the probabilities for the individual sums of spots from the table by counting, see Fig. 12.1.

| Sum of spots | Pairs of numbers (green die, blue die) | Probability |
|---|---|---|
| 2 | (1 \| 1) | 1/36 ≈ 2.8 % |
| 3 | (1 \| 2), (2 \| 1) | 2/36 ≈ 5.6 % |
| 4 | (1 \| 3), (2 \| 2), (3 \| 1) | 3/36 ≈ 8.3 % |
| 5 | (1 \| 4), (2 \| 3), (3 \| 2), (4 \| 1) | 4/36 ≈ 11.1 % |
| 6 | (1 \| 5), (2 \| 4), (3 \| 3), (4 \| 2), (5 \| 1) | 5/36 ≈ 13.9 % |
| 7 | (1 \| 6), (2 \| 5), (3 \| 4), (4 \| 3), (5 \| 2), (6 \| 1) | 6/36 ≈ 16.7 % |
| 8 | (2 \| 6), (3 \| 5), (4 \| 4), (5 \| 3), (6 \| 2) | 5/36 ≈ 13.9 % |
| 9 | (3 \| 6), (4 \| 5), (5 \| 4), (6 \| 3) | 4/36 ≈ 11.1 % |
| 10 | (4 \| 6), (5 \| 5), (6 \| 4) | 3/36 ≈ 8.3 % |
| 11 | (5 \| 6), (6 \| 5) | 2/36 ≈ 5.6 % |
| 12 | (6 \| 6) | 1/36 ≈ 2.8 % |

**Fig. 12.1**  Probability distribution of the sum of spots when rolling two dice

**Fig. 12.2   a, b** Construction
of the histogram for the sum of
spots when rolling two dice

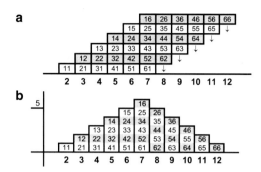

A table, in which the possible sums of spots and the corresponding probabilities are listed, is called a **probability distribution**. The central column in Fig. 12.1 shows which of the combinations of the green and blue dice shown above are related.

This probability distribution can be illustrated in the form of a **histogram**. For this special type of bar chart, rectangles of width 1 are drawn next to each other without gaps; the height of the rectangles corresponds to the probability of the related outcome (here: the related sum of spots).

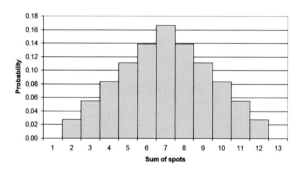

If you omit the brackets and the comma for the pairs of numbers in the central column of Fig. 12.1, you could also draw the graph as shown in Fig. 12.2b. The vertical axis shows the number of possible combinations.

This diagram can also be imagined schematically in this way: For a single dice, there are six possible numbers of spots with the same probability.

| 1 | 2 | 3 | 4 | 5 | 6 |
|---|---|---|---|---|---|

In combination with a second dice, these six cells move 1, 2, 3, 4, 5, 6 units to the right – depending on the number of spots on the second dice (see Fig. 12.2a). Then move the overlaying cells downward onto the horizontal axis (see Fig. 12.2b).

## 12.2   Sums of dice When Rolling Several Regular Hexahedrons

Analogous to the procedure with the rolling of two dice, you can also create a graph showing the distribution of the sum of spots when rolling three dice:

By moving the diagram for the sum of spots when rolling two dice, the histogram for three dice is created (see Fig. 12.3a, b). The diagram shows (in principle) all $6 \cdot 6 \cdot 6 = 216$ possible results of the triple throw.

This idea of shifting (as in Figs. 12.2 and 12.3) can be used to determine the probability distribution of sums of dice, even without recording all the numbers of spots (see the following table with the sum of spots when rolling two dice).

Similarly, this is also possible for sums of spots when throwing more than two dice: Using a spreadsheet and the copy and paste command, you can create the schemes, determine the sums (green colored row) and then calculate the probabilities from these. The row with the number of possibilities is copied and pasted into the next table, shifted by one unit to the right each, etc.

| Sum of spots | 2 | 3 | 4 | 5 | 6 | 7 | 8 | 9 | 10 | 11 | 12 |
|---|---|---|---|---|---|---|---|---|---|---|---|
| | | | | | | 1 | 1 | 1 | 1 | 1 | 1 |
| | | | | | 1 | 1 | 1 | 1 | 1 | 1 | |
| | | | | 1 | 1 | 1 | 1 | 1 | 1 | | |
| | | | 1 | 1 | 1 | 1 | 1 | 1 | | | |
| | | 1 | 1 | 1 | 1 | 1 | 1 | | | | |
| Number of possibilities | 1 | 2 | 3 | 4 | 5 | 6 | 5 | 4 | 3 | 2 | 1 |
| Probability | 0.028 | 0.056 | 0.083 | 0.111 | 0.139 | 0.167 | 0.139 | 0.111 | 0.083 | 0.056 | 0.028 |

| 3 | 4 | 5 | 6 | 7 | 8 | 9 | 10 | 11 | 12 | 13 | 14 | 15 | 16 | 17 | 18 |
|---|---|---|---|---|---|---|---|---|---|---|---|---|---|---|---|
| | | | | | 1 | 2 | 3 | 4 | 5 | 6 | 5 | 4 | 3 | 2 | 1 |
| | | | | 1 | 2 | 3 | 4 | 5 | 6 | 5 | 4 | 3 | 2 | 1 | |
| | | | 1 | 2 | 3 | 4 | 5 | 6 | 5 | 4 | 3 | 2 | 1 | | |
| | | 1 | 2 | 3 | 4 | 5 | 6 | 5 | 4 | 3 | 2 | 1 | | | |
| | 1 | 2 | 3 | 4 | 5 | 6 | 5 | 4 | 3 | 2 | 1 | | | | |
| 1 | 2 | 3 | 4 | 5 | 6 | 5 | 4 | 3 | 2 | 1 | | | | | |
| 1 | 3 | 6 | 10 | 15 | 21 | 25 | 27 | 27 | 25 | 21 | 15 | 10 | 6 | 3 | 1 |
| 0.005 | 0.014 | 0.028 | 0.046 | 0.069 | 0.097 | 0.116 | 0.125 | 0.125 | 0.116 | 0.097 | 0.069 | 0.046 | 0.028 | 0.014 | 0.005 |

In this way, the following histograms of the probability distribution for the sums of spots for rolling two, three, four, and five dice were created.

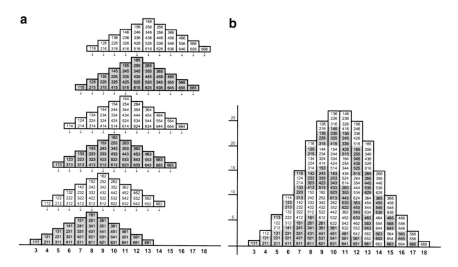

**Fig. 12.3   a, b** Construction of the histogram for the sum of spots when rolling three dice

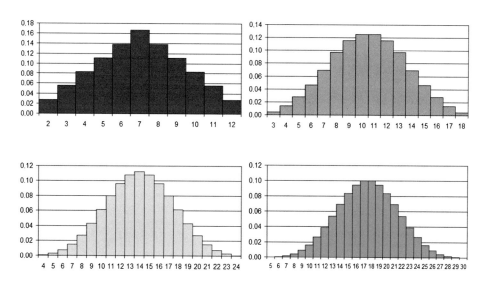

As the number of dice increases, the histograms of the probability distributions for the sum of spots resembles more and more a **bell-shaped curve**. In Fig. 12.4, in addition to the histogram of the sum of dice for rolling five dice, the graph of the so-called **Gaussian density function** of the corresponding **normal distribution** is also shown.

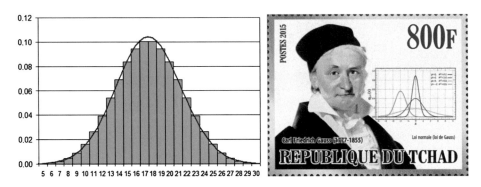

**Fig. 12.4** Histogram of the sum of spots for rolling five dice and graph of the corresponding density function, stamp on C. F. Gauss and the normal distributions (Tchad 2015)

This function has the equation

$$\varphi(x) = \frac{1}{\sigma \cdot \sqrt{2\pi}} \cdot e^{-\frac{1}{2}\left(\frac{x-\mu}{\sigma}\right)^2},$$

where in the case of rolling five dice the parameters $\mu$ and $\sigma$ are $\mu = 17.5$ and $\sigma \approx 3.82$ (see Sect. 12.8).

## 12.3   An Erroneous Notion of Sums of Spots

It is reported that Galileo Galilei (1564–1642) was asked by his ruler whether it would be equally favorable to bet on 6 or 7 as the sum of spots when rolling a die twice.

The reasoning behind this question was: If you roll two *indistinguishable* dice, you cannot tell which of the two dice shows which number of spots. Therefore, only 21 outcomes appear in the central column of the following table, each with the same probability.

| Sum of spots | possible outcomes | Probability | Sum of spots | possible outcomes | Probability |
|---|---|---|---|---|---|
| 2 | {1,1} | 1/21 | 12 | {6,6} | 1/21 |
| 3 | {1,2} | 1/21 | 11 | {5,6} | 1/21 |
| 4 | {1,3}, {2,2} | 2/21 | 10 | {4,6}, {5,5} | 2/21 |
| 5 | {1,4}, {2,3} | 2/21 | 9 | {3,6}, {4,5} | 2/21 |
| 6 | {1,5}, {2,4}, {3,3} | 3/21 | 8 | {2,6}, {3,5}, {4,4} | 3/21 |
| 7 | {1,6}, {2,5}, {3,4} | 3/21 | | | |

The probability distribution in Sect. 12.1, with 36 different results, was based on the assumption that, for example, a green and a blue dice were rolled. However, the color of the dice cannot determine the probability distribution corresponding to the random experiment. If you have two indistinguishable dice, it is no problem to make the two dice distinguishable – for example by puting a tiny mark on one of the dice.

Figures 12.5 and 12.6 show the incorrect and correct probability distribution in contrast.

**Fig. 12.5** Table: Comparing the two "models" of a probability distribution when rolling two dice

| Sum of spots | incorrect probability | correct probability |
|---|---|---|
| 2 | 1/21 = 12/252 | 1/36 = 7/252 |
| 3 | 1/21 = 12/252 | 2/36 = 14/252 |
| 4 | 2/21 = 24/252 | 3/36 = 21/252 |
| 5 | 2/21 = 24/252 | 4/36 = 28/252 |
| 6 | 3/21 = 36/252 | 5/36 = 35/252 |
| 7 | 3/21 = 36/252 | 6/36 = 42/252 |
| 8 | 3/21 = 36/252 | 5/36 = 35/252 |
| 9 | 2/21 = 24/252 | 4/36 = 28/252 |
| 10 | 2/21 = 24/252 | 3/36 = 21/252 |
| 11 | 1/21 = 12/252 | 2/36 = 14/252 |
| 12 | 1/21 = 12/252 | 1/36 = 7/252 |

**Fig. 12.6** Histograms: Comparing the two "models" of a probability distribution when rolling two dice

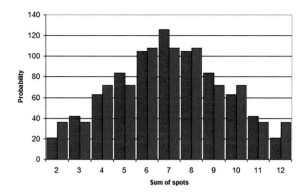

**Suggestions for Reflection and for Investigations**

**A 12.1:** If you compare the two probability distributions in Figs. 12.5 and 12.6, you will realize that the (true) probability for an even and an odd sum of spots is $\frac{1}{2}$. On the other hand, it follows from the wrong model that even sums of spots would be more likely to occur (probability $\frac{12}{21} = \frac{4}{7}$ for an even sum of spots and $\frac{9}{21} = \frac{3}{7}$ for an odd sum of spots). This contradicts the fact, that there are as many even as odd numbers of spots on the dice.

But how would one deal with the following argument? *Even sums of spots occur when the two dice both show an even number or both show an odd number. Odd sum of spots occur when one dice shows an even number and the other an odd number; and since the two dice are indistinguishable, there are fewer cases to consider than with even sums of spots ...*

In the following we are going to analyze whether the error of Galileo's contemporaries could have been detected by an experiment. As explained in **A 12.1** the probability $p_1$ for the sum of spots to be even, was assumed to be $p_1 = \frac{4}{7}$ in the incorrect model, whereas for the correct probability we have $p_2 = \frac{1}{2}$.

For example, if you roll two dice $n = 126$ times, and if you use the wrong model you would expect an even sum of spots to occur in $\frac{4}{7}$ of 126 trials, i.e. 72 of 126 trials of rolling two dice, if you use the correct model you will expect an even sum of spots in half of the trials, that is, in 63 of 126 trials.

**Note**

You calculate the **expected value** $\mu$ of the number of trials with an even sum of dice by calculating the corresponding ratio of the total number of attempts:

$$\frac{4}{7} \text{ from } 126 = \frac{4}{7} \cdot 126 = 72 = \mu_1$$

or

$$\frac{1}{2} \text{ from } 126 = \frac{1}{2} \cdot 126 = 63 = \mu_2.$$

If you roll both dice 126 times, an even sum of spots need not occur *exactly* 72 times or *exactly* 63 times. These numbers, however, are the ones with the highest probability in each models; adjacent values such as 71 and 73 or 62 and 64, however, occur with a similar probability. You can compare the probabilities for certain even sums in Fig. 12.7.

A decision by experiment between the two models could be approached as follows: You define a **critical value** $k$ which, for example, lies in the middle between the two expected values.

If the number of trials with an even sum of spots is greater than $k = \frac{1}{2} \cdot (63 + 72) = 67.5$, then one prefers the (wrong) model with $p_1 = \frac{4}{7}$ as correct, otherwise the model with $p_2 = \frac{1}{2}$.

**Fig. 12.7** Probabilities for certain even sums of spots when rolling two dice 126 times

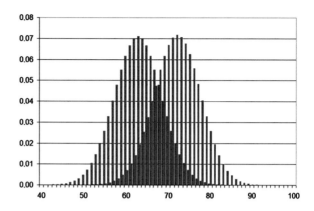

One can calculate that a number of 126 trials would not be great enough to decide with a high degree of certainty between the two models: With a probability of about 21% you would make a wrong decision, because the probability that there are more than 67 trials with an even sum of spots *randomly* is about 21%. Even if one would roll double as often, the probability for a wrong decision (more than 135 times an even sum of spots) would still be about 12%.

A decision-making procedure as described in Sect. 12.10 could also be used.

## 12.4  A Fair Game of Dice

Let's stay with rolling two dice for now and play a simple game where the sum of spots is crucial:

*If the sum of dice is an even number, the first player (player A) wins, if the sum of dice is odd, the second player (player B) wins.*

At first sight (and if you have not studied Sect. 12.3) the game's rule does *not* seem to be fair; because of the possible sums of spots 2, 3, 4, ..., 11, 12 six are even and only five are odd.

This is not what matters, but rather the probability with which such sums of spots occur. Instead of the probabilities, you can also compare the number of possible combinations:

| Sum of spots | even | | | | | | odd | | | | |
|---|---|---|---|---|---|---|---|---|---|---|---|
| | 2 | 4 | 6 | 8 | 10 | 12 | 3 | 5 | 7 | 9 | 11 |
| Number of combinations | 1 | 3 | 5 | 5 | 3 | 1 | 2 | 4 | 6 | 4 | 2 |
| total | 18 possible combinations | | | | | | 18 possible combinations | | | | |

There are many other game rules that are fair: you only have to think about how many ways the number 18 can be represented as a sum of combinations, where the following eleven numbers of combinations are possible summands: 1, 1, 2, 2, 3, 3, 4, 4, 5, 5, 6.

To get 18 as the sum of combinations, you need at least four summands:

$18 = 4 + 4 + 5 + 5$: Player A wins if sum 5, 9, 6, or 8 occurs.

$18 = 2 + 5 + 5 + 6$: Player A wins if sum 3, 6, 8, or 7 occurs or if sum 11, 6, 8, or 7 occurs.

$18 = 3 + 4 + 5 + 6$: This sum includes eight different rules (see **A 12.2** (1)).

Since the remaining seven summands of the possible combinations also add up to 18, the complementary rules of the game are also given, for example:

Player A wins with 5, 9, 6, 8 as the sum of spots $\leftrightarrow$ Player B wins with 2, 3, 4, 7, 10, 11, 12.

---

**Suggestions for Reflection and for Investigations**
**A 12.2:**

1. List the eight possible rules for a fair game where you win with 3, 4, 5, 6 as the numbers of spots.
2. The sum of spots 18 can also be formed from five summands. How many rules can be specified for this?
3. Specify fair rules for three or four players.

---

## 12.5   The Sicherman Dice

In 1978, Martin Gardner reported in his monthly column in *Scientific American,* that the amateur mathematician George Sicherman had discovered another way of labeling the faces of two cubes in such a way that they lead to the same probability distribution of the sum of spots as the usual labeling with the numbers from 1 to 6 (see the following figures of the two corresponding nets of a dice).

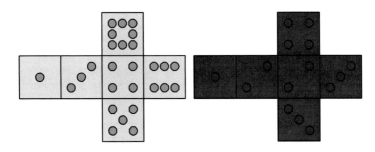

As you can see from the following table of combinations, sum 2 and 12 occur once each, sum 3 and 11 occur twice each, sum 4 and 10 occur three times each, sum 5 and 9 four times each, sum 6 and 8 five times each, and sum 7 six times – just like throwing two ordinary dice:

|   | 1 | 2 | 2 | 3 | 3 | 4 |
|---|---|---|---|---|---|---|
| 1 | 2 | 3 | 3 | 4 | 4 | 5 |
| 3 | 4 | 5 | 5 | 6 | 6 | 7 |
| 4 | 5 | 6 | 6 | 7 | 7 | 8 |
| 5 | 6 | 7 | 7 | 8 | 8 | 9 |
| 6 | 7 | 8 | 8 | 9 | 9 | 10 |
| 8 | 9 | 10 | 10 | 11 | 11 | 12 |

Ordinary dice and the Sichermann dice can also be replaced by wheels of fortune with six sectors of equal size and corresponding labeling:

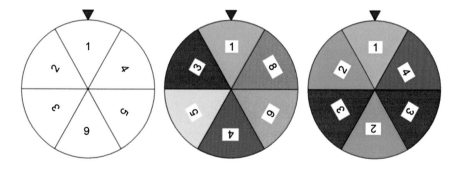

## 12.6   Other Devices with Random Output for Double Throwing

Sicherman's month-long search should not be under-estimated. However, there are another six possibilities, but the amateur mathematician Sicherman did not look for them, because they are not 6-sided dice or substitutes.

For these other cases it is necessary to use appropriately labeled wheels of fortune.

The 36 possible combinations when rolling two ordinary dice can also be represented by a wheel of fortune with 4 sectors ("wheel-of-4") and a wheel of fortune with 9 sectors ("wheel-of-9"). The following figure shows a wheel of 4 whose four 90°-sectors are labeled with the numbers 1, 2, 2, 3, and a wheel-of-9 whose nine 40°-sectors are labeled with the numbers 1, 3, 3, 5, 5, 5, 7, 7, 9. The wheel-of-4 could also be replaced by a regular tetrahedron, which is labeled accordingly; but there is no such substitute for the 9-sided dice.

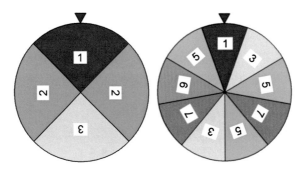

In the corresponding table of combinations you can again read that 2 as the sum of spots occurs once, the sum 3 twice, and so on.

|   | 1 | 3 | 3 | 5 | 5 | 5 | 7 | 7 | 9 |
|---|---|---|---|---|---|---|---|---|---|
| 1 | 2 | 4 | 4 | 6 | 6 | 6 | 8 | 8 | 10 |
| 2 | 3 | 5 | 5 | 7 | 7 | 7 | 9 | 9 | 11 |
| 2 | 3 | 5 | 5 | 7 | 7 | 7 | 9 | 9 | 11 |
| 3 | 4 | 6 | 6 | 8 | 8 | 8 | 10 | 10 | 12 |

But there is an alternative for the combination of a wheel-of-4 and a wheel-of-9 as well: a wheel-of-4 with the labels 1, 4, 4, 7, and a wheel-of-9 with the label 1, 2, 2, 3, 3, 3, 4, 4, 5.

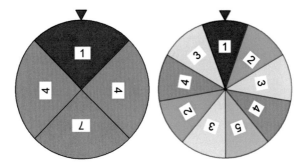

Again, the corresponding table of combinations shows the same probability distribution of the sum of spots as we have with two ordinary 6-sided dice.

|   | 1 | 2 | 2 | 3 | 3 | 3 | 4 | 4 | 5 |
|---|---|---|---|---|---|---|---|---|---|
| 1 | 2 | 3 | 3 | 4 | 4 | 4 | 5 | 5 | 6 |
| 4 | 5 | 6 | 6 | 7 | 7 | 7 | 8 | 8 | 9 |
| 4 | 5 | 6 | 6 | 7 | 7 | 7 | 8 | 8 | 9 |
| 7 | 8 | 9 | 9 | 10 | 10 | 10 | 11 | 11 | 12 |

The 36 possible combinations when rolling two dice can also be represented by a wheel-of-3 and a wheel – of-12. The 120°-sectors of the wheel-of-3 are labeled with the numbers 1, 3, 5 and the 30°-sectors of the wheel-of-12 are labeled with the numbers 1, 2, 2, 3, 3, 4, 4, 5, 5, 5, 6, 6, 7.

The corresponding table of combinations table contains the sums of spots from 2 to 12 with the known frequencies.

|   | 1 | 2 | 2 | 3 | 3 | 4 | 4 | 5 | 5 | 6 | 6 | 7 |
|---|---|---|---|---|---|---|---|---|---|---|---|---|
| 1 | 2 | 3 | 3 | 4 | 4 | 5 | 5 | 6 | 6 | 7 | 7 | 8 |
| 3 | 4 | 5 | 5 | 6 | 6 | 7 | 7 | 8 | 8 | 9 | 9 | 10 |
| 5 | 6 | 7 | 7 | 8 | 8 | 9 | 9 | 10 | 10 | 11 | 11 | 12 |

For this there is also an alternative: the labeling of the wheel-of-3 with 1, 2, 3 and the wheel-of-12 with 1, 2, 3, 4, 4, 5, 5, 6, 6, 7, 8, 9. Here is the corresponding table of combinations:

|   | 1 | 2 | 3 | 4 | 4 | 5 | 5 | 6 | 6 | 7 | 8 | 9 |
|---|---|---|---|---|---|---|---|---|---|---|---|---|
| 1 | 2 | 3 | 4 | 5 | 5 | 6 | 6 | 7 | 7 | 8 | 9 | 10 |
| 2 | 3 | 4 | 5 | 6 | 6 | 7 | 7 | 8 | 8 | 9 | 10 | 11 |
| 3 | 4 | 5 | 6 | 7 | 7 | 8 | 8 | 9 | 9 | 10 | 11 | 12 |

The wheel-of-12 could be replaced by a regular dodecahedron, which is labeled accordingly; a substitute for the wheel-of-3 could be a dice with 1, 1, 2, 2, 3, 3.

Finally, there are two ways to represent the 36 combinations when rolling two dice using a wheel-of-2 and a wheel-of-18. The 180°-sectors of the wheel-of-2, two are labeled with the numbers 1, 2, and the 20°-sectors of the wheel-of-18 are labeled with the

numbers 1, 2, 3, 3, 4, 4, 5, 5, 5, 5, 6, 6, 6, 7, 7, 8, 8, 9, 10 – the corresponding table of
combinations table is as follows:

|   | 1 | 2 | 3 | 3 | 4 | 4 | 5 | 5 | 5 | 6 | 6 | 6 | 7 | 7 | 8 | 8 | 9 | 10 |
|---|---|---|---|---|---|---|---|---|---|---|---|---|---|---|---|---|---|----|
| 1 | 2 | 3 | 4 | 4 | 5 | 5 | 6 | 6 | 6 | 7 | 7 | 7 | 8 | 8 | 9 | 9 | 10 | 11 |
| 2 | 3 | 4 | 5 | 5 | 6 | 6 | 7 | 7 | 7 | 8 | 8 | 8 | 9 | 9 | 10 | 10 | 11 | 12 |

There is also an alternative to this, as can be seen in the following table of combinations:
the labeling of the wheel-of-2 with 1, 4 and the wheel-of-18 with 1, 2, 2, 3, 3, 4, 4, 4, 5,
5, 5, 6, 6, 6, 7, 7, 8.

|   | 1 | 2 | 2 | 3 | 3 | 3 | 4 | 4 | 4 | 5 | 5 | 5 | 6 | 6 | 6 | 7 | 7 | 8 |
|---|---|---|---|---|---|---|---|---|---|---|---|---|---|---|---|---|---|---|
| 1 | 2 | 3 | 3 | 4 | 4 | 4 | 5 | 5 | 5 | 6 | 6 | 6 | 7 | 7 | 7 | 8 | 8 | 9 |
| 4 | 5 | 6 | 6 | 7 | 7 | 7 | 8 | 8 | 8 | 9 | 9 | 9 | 10 | 10 | 10 | 11 | 11 | 12 |

The wheel-of-2 could be replaced by a coin, which is labeled accordingly; there is no
substitute for the 18-sided dice.

## 12.7   Algebraic Background for the Different Display Options

That no other possibilities of representation exist can be proved by means of an ingenious method developed by Leonhard Euler (1707–1783). He considered so-called *generating functions* (this name was given by Pierre-Simon Laplace, 1749–1827):
   The following polynomial corresponds to the labeling of an ordinary dice:

$$f(x) = \tfrac{1}{6} \cdot \left(1x^1 + 1x^2 + 1x^3 + 1x^4 + 1x^5 + 1x^6\right)$$

The exponents represent the numbers of spots occurring, the coefficients the frequency
with which the corresponding numbers are labeled on the dice or wheels, and the factor $\tfrac{1}{6}$
for the probability of each individual number of spots.
   If we look at the product of the generating function with itself (i.e., the square of the
generating function), then we get the generating function for the sum of spots:

$$\begin{aligned} f^2(x) &= \tfrac{1}{36} \cdot \left[1x^1 + 1x^2 + 1x^3 + 1x^4 + 1x^5 + 1x^6\right]^2 \\ &= \tfrac{1}{36} \cdot \left(1x^2 + 2x^3 + 3x^4 + 4x^5 + 5x^6 + 6x^7 + 5x^8 + 4x^9 + 3x^{10} + 2x^{11} + 1x^{12}\right) \end{aligned}$$

Note that now the exponents represent the different possible sums of spots and the coefficients the frequency with which the corresponding sums of spots appear in the table of
combinations, the factor $\tfrac{1}{36}$ is the probability of each of the 36 combinations.
   Such a 12th degree polynomial can be decomposed into factors in different ways.

The best way to start is to factorize the generating function $f$ itself. You get:

$$f(x) = \tfrac{1}{6} \cdot x \cdot (1 + x) \cdot \left(1 + x + x^2\right) \cdot \left(1 - x + x^2\right)$$

So you will find, besides the simplest possibility

$$f^2(x) = \tfrac{1}{6} \cdot \left(1x^1 + 1x^2 + 1x^3 + 1x^4 + 1x^5 + 1x^6\right) \cdot \tfrac{1}{6} \cdot \left(1x^1 + 1x^2 + 1x^3 + 1x^4 + 1x^5 + 1x^6\right),$$

which stands for the usual labeling of the dice, there are another seven ways of factorizing $f^2(x)$. And each of these corresponds to the combination of different devices with random output:

- Alternative labeling of two hexahedrons (Sicherman's model)

$$f^2(x) = \tfrac{1}{36} \cdot \left[x \cdot (1 + x) \cdot \left(1 + x + x^2\right) \cdot \left(1 - x + x^2\right)^2\right] \cdot \left[x \cdot (1 + x) \cdot \left(1 + x + x^2\right)\right]$$
$$= \tfrac{1}{6} \cdot \left[1x + 1x^3 + 1x^4 + 1x^5 + 1x^6 + 1x^8\right] \cdot \tfrac{1}{6} \cdot \left[1x + 2x^2 + 2x^3 + 1x^4\right]$$

- wheel-of-4 and wheel-of-9 (1st possibility)

$$f^2(x) = \tfrac{1}{36} \cdot \left[x \cdot (1 + x)^2\right] \cdot \left[x \cdot \left(1 + x^2 + x^4\right)^2\right]$$
$$= \tfrac{1}{4} \cdot \left[1x + 2x^2 + 1x^3\right] \cdot \tfrac{1}{9} \cdot \left[1x + 2x^3 + 3x^5 + 2x^7 + 1x^9\right]$$

- wheel-of-4 and wheel-of-9 (2nd possibility)

$$f^2(x) = \tfrac{1}{36} \cdot \left[x \cdot \left(1 + x + x^2\right)^2\right] \cdot \left[x \cdot \left(1 + x^3\right)^2\right]$$
$$= \tfrac{1}{9} \cdot \left[1x + 2x^2 + 3x^3 + 2x^4 + 1x^5\right] \cdot \tfrac{1}{4} \cdot \left[1x + 2x^4 + 1x^7\right]$$

- wheel-of-3 and wheel-of-12 (1st possibility)

$$f^2(x) = \tfrac{1}{36} \cdot \left[x \cdot (1 + x)^2 \cdot \left(1 + x^2 + x^4\right)\right] \cdot \left[x \cdot \left(1 + x^2 + x^4\right)\right]$$
$$= \tfrac{1}{12} \cdot \left[1x + 2x^2 + 2x^3 + 2x^4 + 2x^5 + 2x^6 + 1x^7\right] \cdot \tfrac{1}{3} \cdot \left[1x + 1x^3 + 1x^5\right]$$

- wheel-of-3 and wheel-of-12 (2nd possibility)

$$f^2(x) = \tfrac{1}{36} \cdot \left[x \cdot \left(1 + x + x^2\right) \cdot \left(1 + x^3\right)^2\right] \cdot \left[x \cdot \left(1 + x + x^2\right)\right]$$
$$= \tfrac{1}{12} \cdot \left[1x + 1x^2 + 1x^3 + 2x^4 + 2x^5 + 2x^6 + 1x^7 + 1x^8 + 1x^9\right] \cdot \tfrac{1}{3} \cdot \left[1x + 1x^2 + 1x^3\right]$$

- wheel-of-2 and wheel-of-18 (1st possibility)

$$f^2(x) = \tfrac{1}{36} \cdot \left[x \cdot (1 + x) \cdot \left(1 + x^2 + x^4\right)^2\right] \cdot \left[x \cdot (1 + x)\right]$$
$$= \tfrac{1}{18} \cdot \left[1x + 1x^2 + 2x^3 + 2x^4 + 3x^5 + 3x^6 + 2x^7 + 2x^8 + 1x^9 + 1x^{10}\right] \cdot \tfrac{1}{2} \cdot \left[1x + 1x^2\right]$$

- wheel-of-2 and wheel-of-18 (2nd possibility)

$$f^2(x) = \tfrac{1}{36} \cdot \left[x \cdot \left(1 + x + x^2\right)^2 \cdot \left(1 + x^3\right)\right] \cdot \left[x \cdot \left(1 + x^3\right)\right]$$
$$= \tfrac{1}{18} \cdot \left[1x + 2x^2 + 3x^3 + 3x^4 + 3x^5 + 3x^6 + 2x^7 + 1x^8\right] \cdot \tfrac{1}{2} \cdot \left[1x + 1x^4\right]$$

Rolling of two ordinary dice with the usual labeling can, therefore, be replaced by seven alternative random experiments with other devices and with random results that lead to the same probability distribution (see the following overview table in Fig. 12.8).

The factorization of the generating function $f(x) = \frac{1}{6} \cdot x \cdot (1+x) \cdot (1+x+x^2) \cdot (1-x+x^2)$ can be interpreted as saying that the ordinary dice can also be substituted.

The factor $(1-x+x^2)$ that defies factorization can be combined with the others.

- **1st possibility**
  $(1+x) \cdot (1-x+x^2) = 1 - x + x^2 + x - x^2 + x^3 = 1 + x^3$, that is

$$f(x) = \frac{1}{6} \cdot x \cdot (1+x^3) \cdot (1+x+x^2)$$

The product can be interpreted in two ways with regard to the labeling of devices with random output:

$$f(x) = \frac{1}{2} \cdot (x^1 + x^4) \cdot \frac{1}{3} \cdot (x^0 + x^1 + x^2)$$

or

$$f(x) = \frac{1}{2} \cdot (x^0 + x^3) \cdot \frac{1}{3} \cdot (x^1 + x^2 + x^3).$$

And these represent combinations of a wheel-of-2, labeled with the numbers 1 and 4, and a wheel-of-3, labeled with the numbers 0, 1, and 2, or as the combination of a wheel-of-2, labeled with the numbers 0 and 3, and a wheel-of-3, labeled with the numbers 1, 2, and 3 (see the two following tables of combinations).

| 1st wheel of fortune: Number of sectors | Labeling | 2nd wheel of fortune: Number of sectors | Labeling |
|---|---|---|---|
| 6 (hexahedron) | (1, 3, 4, 5, 6, 8) | 6 (hexahedron) | (1, 2, 2, 3, 3, 4) |
| 4 (tetrahedron) | (1, 2, 2, 3) | 9 | (1, 3, 3, 5, 5, 5, 7, 7, 9) |
| 4 (tetrahedron) | (1, 4, 4, 7) | 9 | (1, 2, 2, 3, 3, 3, 4, 4, 5) |
| 3 | (1, 3, 5) | 12 (dodecahedron) | (1, 2, 2, 3, 3, 4, 4, 5, 5, 6, 6, 7) |
| 3 | (1, 2, 3) | 12 (dodecahedron) | (1, 2, 3, 4, 4, 5, 5, 6, 6, 7, 8, 9) |
| 2 (coin) | (1, 2) | 18 | (1, 2, 3, 3, 4, 4, 5, 5, 5, 6, 6, 6, 7, 7, 8, 8, 9, 10) |
| 2 (coin) | (1, 4) | 18 | (1, 2, 2, 3, 3, 3, 4, 4, 4, 5, 5, 5, 6, 6, 6, 7, 7, 8) |

**Fig. 12.8** Summary: Possible devices to substitute two ordinary labeled dices

| | 0 | 1 | 2 |
|---|---|---|---|
| 1 | 1 | 2 | 3 |
| 4 | 4 | 5 | 6 |

| | 1 | 2 | 3 |
|---|---|---|---|
| 0 | 1 | 2 | 3 |
| 3 | 4 | 5 | 6 |

- **2nd possibility**

$\left(1 + x + x^2\right) \cdot \left(1 - x + x^2\right) = 1 - x + x^2 + x - x^2 + x^3 + x^2 - x^3 + x^4 = 1 + x^2 + x^4$, that is

$f(x) = \frac{1}{6} \cdot x \cdot (1 + x) \cdot \left(1 + x^2 + x^4\right)$ and from this we have

$$f(x) = \frac{1}{2} \cdot (x^1 + x^2) \cdot \frac{1}{3} \cdot \left(x^0 + x^2 + x^4\right)$$

or

$$f(x) = \frac{1}{2} \cdot (x^0 + x^1) \cdot \frac{1}{3} \cdot \left(x^1 + x^3 + x^5\right).$$

These products can be interpreted as the combination of a wheel-of-2, labeled with the numbers 1 and 2, and a wheel-of-3, labeled with the numbers 0, 2, and 4, or as the combination of a wheel-of-2, labeled with the numbers 0 and 1, and a wheel-of-3, labeled with the numbers 1, 3, and 5 (see the two corresponding tables of combinations).

| | 0 | 2 | 4 |
|---|---|---|---|
| 1 | 1 | 3 | 5 |
| 2 | 2 | 4 | 6 |

| | 1 | 3 | 5 |
|---|---|---|---|
| 0 | 1 | 3 | 5 |
| 1 | 2 | 4 | 6 |

## 12.8   Probability Distribution of Sums of Spots for Rolling *n* Dice

The method of generating functions provides an algebraic procedure for determining the probability distribution of the sum of spots for rolling *n* dice.

By multiplying (rather laboriously) or with the help of a computer algebra system (CAS) you can find out that, for example the following is valid:

$$\begin{aligned} f^3(x) &= \tfrac{1}{216} \cdot \left[1x^1 + 1x^2 + 1x^3 + 1x^4 + 1x^5 + 1x^6\right]^3 \\ &= \tfrac{1}{216} \cdot \left(1x^3 + 3x^4 + 6x^5 + 10x^6 + 15x^7 + 21x^8 + 25x^9 + 27x^{10}\right. \\ &\quad \left. + 27x^{11} + 25x^{12} + 21x^{13} + 15x^{14} + 10x^{15} + 6x^{16} + 3x^{17} + 1x^{18}\right) \end{aligned}$$

$$f^4(x) = \tfrac{1}{1296} \cdot \left[1x^1 + 1x^2 + 1x^3 + 1x^4 + 1x^5 + 1x^6\right]^4$$
$$= \tfrac{1}{1296} \cdot \big(1x^4 + 4x^5 + 10x^6 + 20x^7 + 35x^8 + 56x^9$$
$$+ 80x^{10} + 104x^{11} + 125x^{12} + 140x^{13} + 146x^{14}$$
$$+ 140x^{15} + 125x^{16} + 104x^{17} + 80x^{18} + 56x^{19}$$
$$+ 35x^{20} + 20x^{21} + 10x^{22} + 4x^{23} + 1x^{24}\big)$$

The method of the generating functions proves to be elegant also because with its help the expected value $\mu = E(X)$ and the standard deviation $\sigma$, or the variance $V(X)$ for sums of spots can be easily determined.

If a probability distribution is given by a generating function, then the following applies:

$$\mu = E(X) = f'(1); \quad \sigma^2 = V(X) = f'(1) + f''(1) - f'(1)^2,$$
so here: $f'(x) = \tfrac{1}{6} \cdot \left(1 + 2x + 3x^2 + 4x^3 + 5x^4 + 6x^5\right)$ with

$$f'(1) = \tfrac{1}{6} \cdot (1 + 2 + 3 + 4 + 5 + 6) = \tfrac{21}{6} = 3.5 = \mu$$

and

$$f''(x) = \tfrac{1}{6} \cdot \left(2 + 6x + 12x^2 + 20x^3 + 30x^4\right) \text{ with}$$

$$f''(1) = \tfrac{1}{6} \cdot (2 + 6 + 12 + 20 + 30) = \tfrac{70}{6} = \tfrac{35}{3} = 11.\overline{6},$$

$$\text{so } \sigma^2 = \tfrac{7}{2} + \tfrac{35}{3} - \tfrac{49}{4} = \tfrac{35}{12} = 2.91\overline{6}.$$

Because of the linearity of expected value and variance, there are corresponding multiples of this for the sums of spots of two or more throws (see the following table).

| 2-fold cast | 3-fold cast | 4-fold cast | 5-fold cast |
|---|---|---|---|
| $\mu = 7$ | $\mu = 10.5$ | $\mu = 14$ | $\mu = 17.5$ |
| $\sigma^2 = \tfrac{35}{6} = 5.8\overline{3}$ | $\sigma^2 = \tfrac{35}{4} = 8.75$ | $\sigma^2 = \tfrac{35}{3} = 11.\overline{6}$ | $\sigma^2 = \tfrac{175}{12} = 14.58\overline{3}$ |

The formulas for the expected value and the variance can also derived in an *elementary* way. If generally you look at a wheel of fortune with $m$ sectors of equal size, which are labeled with the numbers 1, 2, 3, ..., $m$ ("number of spots"), then the expected value of the number of spots as defined applies as follows:

$$\mu = \tfrac{1}{m} \cdot (1 + 2 + 3 + \ldots + m)$$

For this purpose, one can write, calculating according to the formula (2.1):

$$\mu = \tfrac{1}{m} \cdot \tfrac{m \cdot (m+1)}{2} = \tfrac{m+1}{2}$$

The following applies to the variance:

$$\sigma^2 = \frac{1}{m} \cdot \left(1^2 + 2^2 + 3^2 + \ldots + m^2\right) - \left(\frac{m+1}{2}\right)^2$$

From this it follows according to the formula **(2.4)**:

$$\sigma^2 = \frac{1}{m} \cdot \frac{m \cdot (m+1) \cdot (2m+1)}{6} - \left(\frac{m+1}{2}\right)^2 = \frac{m^2-1}{12}$$

---

**Formula**

Given is a wheel of fortune with $m$ sectors of equal size, which are labeled with the numbers 1, 2, 3, ..., $m$. The following applies.

**For the expected value $\mu$ of the randomly generated numbers:**

$$\mu = \frac{m+1}{2} \tag{12.1}$$

**For the variance V(X) and the standard deviation $\sigma$ of the randomly generated numbers**

$$\sigma^2 = \frac{m^2-1}{12}, \text{ so } \sigma = \sqrt{\frac{m^2-1}{12}} \tag{12.2}$$

---

## 12.9   Probability Distributions of the Platonic solids

Among the dice that you nowadays find in toy shops are not only the cube but also the other platonic solids: Tetrahedron (*tetra,* Eng. four), octahedron (*octo,* Eng. eight), dodecahedron (*dodeka,* Eng. twelve), icosahedron (*eikosa,* Eng. twenty). These are usually labeled according to their number of cells with the natural numbers from 1 to 4 or 8 or 12 or 20 ("number of spots").

Source: https://commons.wikimedia.org/wiki/File:BluePlatonicDice.jpg

All previous considerations for dice (hexahedra) can also be transferred to the other platonic solids. With the help of the formulas **(2.1)** and **(2.2)** the expected value and the standard deviation of the number of spots on platonic bodies can be calculated in an elementary way (see the following table). Because of the linearity, one receives value of the expectation and standard deviation of sums of spots by corresponding multiples or sums from this table.

| | Tetra-hedron | Hexa-hedron | Octa-hedron | Dodeca-hedron | Icosa-hedron |
|---|---|---|---|---|---|
| Number $m$ of faces | 4 | 6 | 8 | 12 | 20 |
| Expected value $\mu$ | 2.5 | 3.5 | 4.5 | 6.5 | 10.5 |
| Standard-deviation $\sigma$ | $\sqrt{\frac{15}{12}} \approx 1.12$ | $\sqrt{\frac{35}{12}} \approx 1.71$ | $\sqrt{\frac{63}{12}} \approx 2.29$ | $\sqrt{\frac{143}{12}} \approx 3.45$ | $\sqrt{\frac{399}{12}} \approx 5.77$ |

**Suggestions for Reflection and for Investigations**
**A 12.3:** A regular tetrahedron is labeled with the numbers 1, 2, 3, 4.

1. Examine which sums of spots can occur when rolling two tetrahedra (table of combinations, histogram).
2. Determine the generating function for the numbers of spots of a tetrahedron, and use it to determine wheels of fortune that could replace a tetrahedron.
3. Determine possible factorizations of the term of the generating function for the sums when rolling two tetrahedral, and use them to determine three alternative devices with random output that can be used to substitute the experiment with tetrahedra.
4. Use the generating function to determine the expected value and standard deviation of the number of spots of a tetrahedron.

## Sums of Spots When Rolling Different Platonic Solids
Instead of rolling two dice (or other platonic solids) you can examine the sum of spots when rolling different platonic solids. The procedure is the same.

**Example: Sum of Spots When Rolling a Tetrahedron and a Hexahedron**
As above, we first determine the table of combinations, the probability distribution, and the corresponding histogram. For the histogram, the trapezoidal shape is striking (different from the triangular shape of the sum of two identical dice, see Sect. 12.1).

| | 1 | 2 | 3 | 4 | 5 | 6 |
|---|---|---|---|---|---|---|
| **1** | 2 | 3 | 4 | 5 | 6 | 7 |
| **2** | 3 | 4 | 5 | 6 | 7 | 8 |
| **3** | 4 | 5 | 6 | 7 | 8 | 9 |
| **4** | 5 | 6 | 7 | 8 | 9 | 10 |

| Sum of spots | 2 | 3 | 4 | 5 | 6 | 7 | 8 | 9 | 10 |
|---|---|---|---|---|---|---|---|---|---|
| Probability | $\frac{1}{24}$ | $\frac{2}{24}$ | $\frac{3}{24}$ | $\frac{4}{24}$ | $\frac{4}{24}$ | $\frac{4}{24}$ | $\frac{3}{24}$ | $\frac{2}{24}$ | $\frac{1}{24}$ |

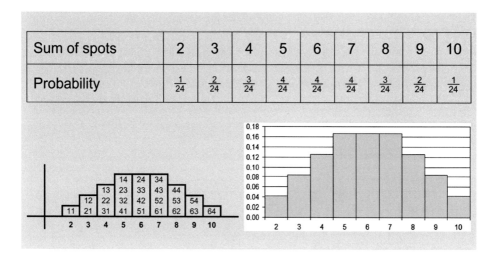

The probability distribution can also be determined by using the two generating functions:

$$\tfrac{1}{4} \cdot \left(1x^1 + 1x^2 + 1x^3 + 1x^4\right) \cdot \tfrac{1}{6} \cdot \left(1x^1 + 1x^2 + 1x^3 + 1x^4 + 1x^5 + 1x^6\right)$$
$$= \tfrac{1}{24} \cdot \left(1x^2 + 2x^3 + 3x^4 + 4x^5 + 4x^6 + 4x^7 + 3x^8 + 2x^9 + 1x^{10}\right)$$

**Suggestions for Reflection and for Investigations**
**A 12.4:** Determine the possible decompositions for the generating function of the sum of spots of tetrahedron and hexahedron, and from these determine the eight alternative combinations of devices with random output with the same probability distribution.

## 12.10   Comparison of Probability Distributions with Equal Sums of Spots

Random experiments with two different platonic solids are of particular interest, where the possible sums of spots may be the same as in experiments with two identical solids.
  There are three possibilities for this:

- When two hexahedra are rolled or a tetrahedron and an octahedron, the values 2, 3, 4, …, 12 can occur as sums of spots.
- When two octahedra are rolled, or a tetrahedron and a dodecahedron, the values 2, 3, 4, …, 16 can occur as sums of spots.
- When two dodecahedra are rolled or a tetrahedron and an icosahedron, the values 2, 3, 4, …, 24 can occur as sums of spots.

The histograms in Fig. 12.9a–c show the different probability distributions. We have already pointed out the difference in shape (triangular or trapezoidal): sums of spots which are close to the expected value have a higher probability in a double throw with the *same* polyhedron, whereas sums of spots at the "ends" of the distribution are more likely to occur when two *different* polyhedra are thrown.

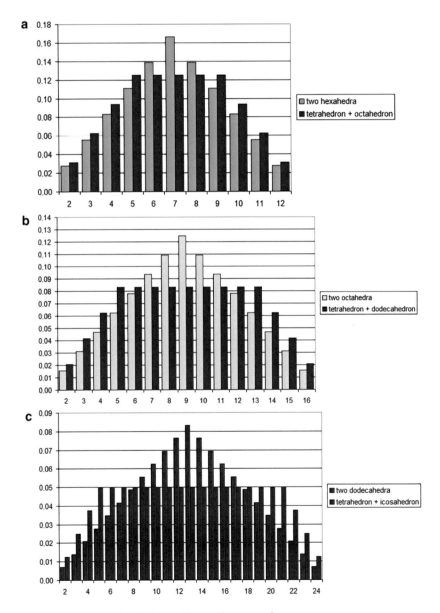

**Fig. 12.9   a–c** Probability distributions with matching sums of spots

**A Game: Would you Place a Bet on the Question Which Combination of Polyhedra was Used?**

What do you think about the following game: The gamemaster rolls the chosen combination of polyhedra (e.g.) 5 times and each time he announces what sums of spots appear. The gambler bets on which of the two combinations of dice were rolled.

As the player can calculate the probabilities with which the sums of spots occur, he or she can determine the chances for which of the two possible combinations is correct.

> **Example**
> The gamemaster announces the following sums of spots: 9, 5, 10, 8, 6.
>
> $$P_{\text{two hexahedron}}(9,\ 5,\ 10,\ 8,\ 6) = \frac{4}{36} \cdot \frac{4}{36} \cdot \frac{3}{36} \cdot \frac{5}{36} \cdot \frac{5}{36} \approx 1.985 \cdot 10^{-5}.$$
>
> $$P_{\text{tetrahedron + octahedron}}(9,\ 5,\ 10,\ 8,\ 6) = \frac{4}{32} \cdot \frac{4}{32} \cdot \frac{3}{32} \cdot \frac{4}{32} \cdot \frac{4}{32} \approx 2.289 \cdot 10^{-5}.$$
>
> Since this sequence of sums has a higher probability when rolling tetrahedron and octahedron than when rolling two hexahedron, one would choose the second combination.
>
> However, the difference between the two probabilities is not very large (odds ratio: 46.4–53.6%).

## 12.11   An Example of the Central Limit Theorem

The **central limit theorem** of probability and statistics states that under certain conditions the sum of several random variables are approximately normally distributed and that the closeness of the approximation by a normal distribution increases with the number of summands.

The following figure shows the probability distribution when throwing one of each of the three of the platonic solids, namely the tetrahedron, hexahedron, and octahedron.

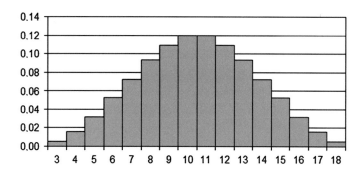

In the next figure a dodecahedron is added:

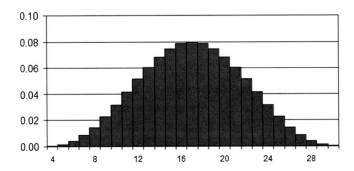

If you throw all five regular polyhedra, the histogram of the sum of spots has almost the shape of a bell curve. In the function term of the density function $\varphi$ with

$$\varphi(x) = \frac{1}{\sigma \cdot \sqrt{2\pi}} \cdot e^{-\frac{1}{2}\cdot\left(\frac{x-\mu}{\sigma}\right)^2}$$

the parameters $\mu$ and $\sigma$ appear with.

$\mu = 2.5 + 3.5 + 4.5 + 6.5 + 10.5 = 27.5$ and

$\sigma^2 = \frac{15}{12} + \frac{35}{12} + \frac{63}{12} + \frac{143}{12} + \frac{399}{12} = \frac{655}{12} = 54.8\overline{3}$, so $\sigma \approx 7.39$.

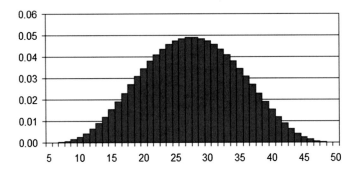

The expected values and standard deviations can be used to make predictions for sums of spots in multiple trials. If such random tests are carried out very often, the possible sums of spots are approximately normally distributed and one can use the so-called **sigma rules.** These are rules of thumb about the probability for intervals lying symmetrically to the expected value. They apply exactly to normal distributions, but approximately to the sums of spots when several dice are rolled.

---

**Rule**

**Sigma rules for sums of spots**

If the sum of spots from multiply rolling of dice can be described approximately by a normal distribution with expected value $\mu$ and standard deviation $\sigma$, then this sum of spots lies

- in about two thirds of the experiments in the interval between $\mu - \sigma$ and $\mu + \sigma$,
- with a probability of about 90% in the interval between $\mu - 1.64\sigma$ and $\mu + 1.64\sigma$,
- with a probability of about 95% in the interval between $\mu - 1.96\sigma$ and $\mu + 1.96\sigma$,
- with a probability of about 99% in the interval between $\mu - 2.58\sigma$ and $\mu + 2.58\sigma$.

**Example: Sum of Spots When Rolling the Five Platonic Solids**
Since the expected value $\mu = 27.5$ and the standard deviation is $\sigma \approx 7.39$, the following predictions are possible:

- In about two thirds of the experiments the sum of spots is between 21 and 34.
- In nine out of ten experiments the sum of spots is between 16 and 39.
- In 19 of 20 experiments the sum of spots is between 14 and 41.
- Only in 1% of the experiments the sum of spots is less than 9 or greater than 46.

The exact probabilities (rounded to one decimal place) are
(1) 62.7% (2) 89.6% (3) 95.1% (4) 0.2%
It turns out that the probability for such extremely small or large sums of spots is even smaller, so the rule of thumb is rather imprecise here.

**Suggestions for Reflection and for Investigations**
**A 12.5:** The sum of spots, when rolling a hexahedron [octahedron] ten times, can be considered normally distributed.
Formulate predictions – similar to the example above – regarding the intervals in which the sums of spots will lie.

## 12.12  Determining Sums of Dice Using Markov Chains

In the theory of Markov chains one deals with **states** of a system and the **transitions** between the states. The sums of dice spots can be interpreted as states.

If, for example, a regular hexahedron is rolled for the first time, then with a probability of $\frac{1}{6}$ a transition occurs from state 0 to state 1, 2, 3, 4, 5, or 6 – depending on the number of spots.

With the second throw transitions to the states 2, 3, 4, …, 12 are possible all transitions happen with the probability $\frac{1}{6}$. Assuming these conditions, then

- the transition to state 2 is only possible from state 1
- the transition to state 3 from both state 1 and state 2

- the transition to state 4 from states 1, 2, 3,
- the transition to state 5 from the states 1, 2, 3, 4
- the transition to state 6 from the states 1, 2, 3, 4, 5
- the transition to state 7 from the states 1, 2, 3, 4, 5, 6
- the transition to state 8 is no longer possible from state 1, but from states 2, 3, 4, 5, 6

State 7 was not yet been reached after the first throw, which means it has the probability 0, and so state 7 can also be mentioned in the list of possible accesses without this having any influence. So we can write the last line as follows:

- the transition to state 8 is possible from states 2, 3, 4, 5, 6, 7

(Mentioning that state 7 has the advantage that six preceding states are used, so we can go on further, analogously).

- the transition to state 9 from the states 3, 4, 5, 6, 7, 8
- ...
- the transition to state 12 from the states 6, 7, 8, 9, 10, 11.

The calculation of the probability $P(n, k)$, that the system is in state $k$ after $n$ throws can be calculated recursively from the probabilities of the previous steps:

$$P(2,2) = \tfrac{1}{6} \cdot P(1,1)$$
$$P(2,3) = \tfrac{1}{6} \cdot [P(1,1) + P(1,2)]$$
$$P(2,4) = \tfrac{1}{6} \cdot [P(1,1) + P(1,2) + P(1,3)]$$
$$P(2,5) = \tfrac{1}{6} \cdot [P(1,1) + P(1,2) + P(1,3) + P(1,4)]$$
$$P(2,6) = \tfrac{1}{6} \cdot [P(1,1) + P(1,2) + P(1,3) + P(1,4) + P(1,5)]$$
$$P(2,7) = \tfrac{1}{6} \cdot [P(1,1) + P(1,2) + P(1,3) + P(1,4) + P(1,5) + P(1,6)]$$
$$P(2,8) = \tfrac{1}{6} \cdot [P(1,2) + P(1,3) + P(1,4) + P(1,5) + P(1,6) + P(1,7)]$$

...

$$P(2,11) = \tfrac{1}{6} \cdot [P(1,5) + P(1,6) + P(1,7) + P(1,8) + P(1,9) + P(1,10)]$$

$$P(2,12) = \tfrac{1}{6} \cdot [P(1,6) + P(1,7) + P(1,8) + P(1,9) + P(1,10) + P(1,11)]$$

At the transition to the 3rd step you have to increase the two parameter values $n$ and $k$ up by 1 each time, so:

$$P(3,3) = \tfrac{1}{6} \cdot P(2,2)$$

$$P(3,4) = \tfrac{1}{6} \cdot [P(2,2) + P(2,3)]$$

$$P(3,5) = \tfrac{1}{6} \cdot [P(2,2) + P(2,3) + P(2,4)]$$

...

$$P(3,8) = \tfrac{1}{6} \cdot [P(2,2) + P(2,3) + P(2,4) + P(2,5) + P(2,6) + P(2,7)]$$

...

$$P(3,13) = \tfrac{1}{6} \cdot [P(2,7) + P(2,8) + P(2,9) + P(2,10) + P(2,11) + P(2,12)]$$

And then one has to continue this until the next greatest possible state:

$$P(3,18) = \tfrac{1}{6} \cdot [P(2,12) + P(2,13) + P(2,14) + P(2,15) + P(2,16) + P(2,17)]$$

In Excel © the first part can be realized with the simple copy and paste command, the second part with drag and drop (see the following graphic).

| n\k | 1 | 2 | 3 | 4 | 5 | 6 | 7 | 8 | 9 | 10 | 11 | 12 | 13 |
|---|---|---|---|---|---|---|---|---|---|---|---|---|---|
| 1 | 0.167 | 0.167 | 0.167 | 0.167 | 0.167 | 0.167 | 0 | 0 | 0 | 0 | 0 | 0 | 0 |
| 2 | | 0.028 | 0.056 | 0.083 | 0.111 | 0.139 | 0.167 | 0.139 | 0.111 | 0.083 | 0.056 | 0.028 | |
| 3 | | | 0.005 | 0.014 | 0.028 | 0.046 | 0.069 | 0.097 | 0.116 | 0.125 | 0.125 | 0.116 | 0.097 |
| 4 | | | | 0.001 | 0.003 | 0.008 | 0.015 | 0.027 | 0.043 | 0.062 | 0.080 | 0.096 | 0.108 |
| 5 | | | | | 0.000 | 0.001 | 0.002 | 0.005 | 0.009 | 0.016 | 0.026 | 0.039 | 0.054 |

One could standardize the recursion formulas in such a way that terms with six summands are used everywhere. To do this, one would have to introduce states with the numbers $0, 1, -1, -2, -3, -4$ in the first step and assign the probability 0 to these states.

Then the uniform recursion formula for $n \geq 2$ and $k$ with $n \leq k \leq 6n$ is as follows:

$$P(n,k) = \tfrac{1}{6} \cdot [P(n-1, k-6) + P(n-1, k-5) + P(n-1, k-4)$$
$$+ P(n-1, k-3) + P(n-1, k-2) + P(n-1, k-1)]$$

with $P(1,1) = P(1,2) = P(1,3) = P(1,4) = P(1,5) = P(1,6) = \tfrac{1}{6}$.
and $P(n,k) = 0$ for $k < n$.

From the rows of the following table you can read the probability of achieving the sum $k$ of spots when rolling $n$ dice.

| n\k | -4 | -3 | -2 | -1 | 0 | 1 | 2 | 3 | 4 | 5 | 6 | 7 | 8 |
|---|---|---|---|---|---|---|---|---|---|---|---|---|---|
| 1 | 0 | 0 | 0 | 0 | 0 | 0.167 | 0.167 | 0.167 | 0.167 | 0.167 | 0.167 | 0 | 0 |
| 2 | | 0 | 0 | 0 | 0 | 0 | 0.028 | 0.056 | 0.083 | 0.111 | 0.139 | 0.167 | 0.139 |
| 3 | | | 0 | 0 | 0 | 0 | 0 | 0.005 | 0.014 | 0.028 | 0.046 | 0.069 | 0.097 |
| 4 | | | | 0 | 0 | 0 | 0 | 0 | 0.001 | 0.003 | 0.008 | 0.015 | 0.027 |
| 5 | | | | | 0 | 0 | 0 | 0 | 0 | 0.000 | 0.001 | 0.002 | 0.005 |

## 12.13  References to Further Literature

On **Wikipedia** you can find further information and literature on the keywords in English (German, French):

- Dice (Spielwürfel, Dé)
- Dice games (Würfelspiel, Jeu de dés)
- Histogram (Histogramm, Histogramme)
- Expectation value (Erwartungswert, Espérance)
- Standard deviation (Standardabweichung, Écart type)
- Probability-generating function (Wahrscheinlichkeitserzeugende Funktion, Fonction génératrice des probabilités)
- Normal distribution (Normalverteilung, Loi normale*)
- Central limit theorem (Zentraler Grenzwertsatz, Théorème central limite)

*) *Marked as an article worth reading.*

Extensive informations can be found at **Wolfram Mathworld** under the keywords:

- Dice, Sicherman dice, Dice games, Expectation value, Standard deviation, Normal distribution.

# The Missing Square

# 13

*There are things which seem incredible to most men who have not studied Mathematics.*

*(Archimedes, Greek mathematician and physicist, 287–212 B.C.)*

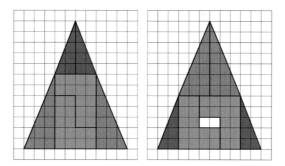

In 1858 Sam Loyd (1841–1911), the inventor and collector of mathematical riddles, was the first to present the problem of the missing square. These riddles can be found in Sect. 13.5 of this chapter.

In the 1950s, Martin Gardner (1914–2010) reported in his monthly column in *Scientific American* that the magician Paul Curry (1917–1986) showed the paradox of two missing squares in his show (see the figures above). Since then the puzzle has also been called *Curry's triangle paradox*".

But also numerous other collections of mathematical puzzles contain explanations of this paradox.

© Springer-Verlag GmbH Germany, part of Springer Nature 2021
H. K. Strick, *Mathematics is Beautiful*, https://doi.org/10.1007/978-3-662-62689-4_13

## 13.1   Apparently Congruent Figures

**Puzzle 1**

In the following figure on the left you can see a square (yellow) with the side length 13. A right-angled triangle (green) is attached to it, whose legs have the side lengths 13 and 21, and above it a right-angled triangle (light blue), whose legs have the side lengths 13 and 8. The figure on the right shows a rectangle (orange) with the side lengths 8 and 21 and the same right-angled triangles as in the figure on the left, this time arranged in interchanged positions.

The two figures seem to be congruent. When checking the areas again, however, it is striking that the area of the orange rectangle in the figure on the right is $8 \cdot 21 = 168$ units and so 1 unit smaller than the yellow square in the figure on the left: $13 \cdot 13 = 169$ units.

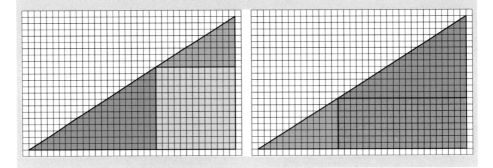

Before you try to get to the bottom of puzzle 1, you should look at a similar problem in the following figures:

**Puzzle 2**

The legs of the light-blue triangle have the side lengths 5 and 8 and those of the green triangle have the side lengths 8 and 13. Again in this puzzle the two figures seem to be congruent.

When recalculating, it is amazing again that the areas of rectangle and square do not match: The area of the rectangle is $5 \cdot 13 = 65$ units and it is 1 unit greater than that of the square with $8 \cdot 8 = 64$ units.

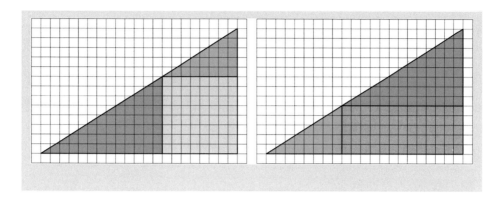

Only when considering the following figures of puzzle 3 one has perhaps a chance to find out what is wrong here.

**Puzzle 3**

The legs of the light blue and the green triangle have the side lengths 3 and 5 and 5 and 8. And in this example the two figures seem to be congruent at first sight.

Here, too, it is intriguing that the areas of the two figures differ by 1 unit: The area of the rectangle is $3 \cdot 8 = 24$ and the area of the square is $5 \cdot 5 = 25$ units.

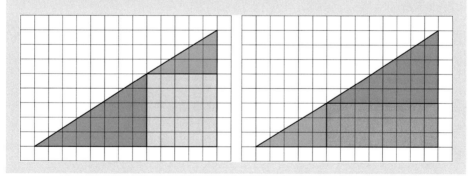

If you look *very closely* you will (perhaps) realize that the hypotenuses of the blue and green triangle do not form a continuous line.

This "kink" in the line of sight becomes even more obvious when the figure shown on the right in the last illustration is doubled. Now you can see a narrow white band (more precisely: a parallelogram). And this has exactly an area of 1 unit.

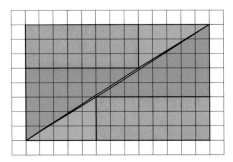

This is the missing square unit!

That is because the whole figure is – measured outside – a rectangle with the area $(3+5) \cdot (5+8) = 8 \cdot 13 = 104$ units.

It consists of two rectangles of $3 \cdot 8 = 24$ units so together 48 units, the two green triangles ($5 \cdot 8 = 40$ units) and the two light blue triangles ($3 \cdot 5 = 15$ units), a total of 103 units. The white band can only be discovered because the whole figure is presented with enlarged length units.

This is not so obvious in the figures in the other illustrations: The white band is hardly visible in the figure with the doubled yellow $8 \times 8$ square from puzzle 2. Here, the band also has an area of 1 unit, but the total figure has a larger area. And it is quite impossible to spot the white band with an area of 1 unit in the doubled figure belonging to puzzle 1.

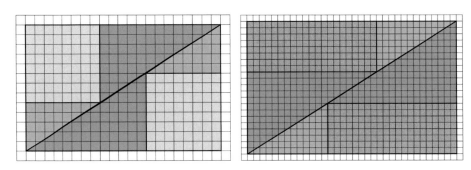

**Suggestions for Reflection and for Investigations**
**A 13.1:** Determine the area of the partial figures in the last two illustrations and the total area each. What is the area of the white band?

The paradox of the missing square is thus related to the fact that the light blue and the green triangle have different *slopes*. However, the difference is so small that it is not apparent. By the way, the side lengths we have chosen in the figures are numbers

Fibonacci's sequence (see Chap. 3). The magic trick of the missing square works because two properties accumulate here:

- The ratio of two consecutive numbers in the Fibonacci sequence very quickly approaches the limit $\frac{\sqrt{5}-1}{2} \approx 0.618\ldots$ – and therefore the gradient angles of the triangles differ less and less from each other (see Tab. in A 13.2).
- The square of an element of the Fibonacci sequence always differs from the product of its two neighbors by 1, namely it is alternately larger by 1 and smaller by 1:

$$2^2 = 1 \cdot 3 + 1; \quad 3^2 = 2 \cdot 5 - 1; \quad 5^2 = 3 \cdot 8 + 1; \quad 8^2 = 5 \cdot 13 - 1; \quad 13^2 = 8 \cdot 21 + 1; \quad 21^2 = 13 \cdot 34 - 1; \quad \ldots$$

This property was discovered by the Italian mathematician and astronomer Giovanni Domenico Cassini (1625–1712), among others, and is called **Cassini's identity** in his honor.

(*Hint:* The proof of the formula can be done, for example, by mathematical induction).

---

**Formula**

**Cassini's identity**
For three consecutive numbers $f_{n-1}, f_n, f_{n+1} (n \in \mathbb{N})$ of the Fibonacci sequence, the following equation applies:

$$f_{n-1} \cdot f_{n+1} = f_n^2 + (-1)^n \tag{13.1}$$

In the figures in puzzle 1 the green triangle has a gradient angle of $\tan^{-1}\left(\frac{13}{21}\right) \approx 31.76°$, the light blue of $\tan^{-1}\left(\frac{8}{13}\right) \approx 31.61°$. This difference of 0.15° can hardly be seen! The optical illusion is reasonable because the whole figure is seen as a triangle, whose gradient angle is determined by the ratio $(13 + 8):(21 + 13) = 21:34$ and so results in: $\tan^{-1}\left(\frac{21}{34}\right) \approx 31.70°$. ◀

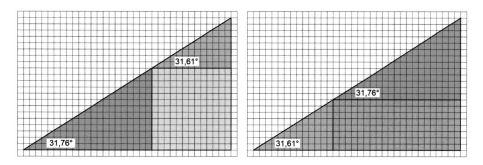

But in fact the figure on the left has a "kink" to the outside, the figure on the right has a kink to the inside. Because of the small difference in the gradient angle, you cannot see the inward kink even when doubling the figure on the right and, therefore, you cannot see the white band.

The difference of 1 unit can be made visible not only by calculation, but also by dissection. The two initial yellow figures, that is, the square and the rectangle, can be tessellated as follows: In puzzle 3, the $5 \times 5$ square can be covered by three puzzle pieces; one of the pieces protrudes with 1 unit from the $3 \times 8$ rectangle.

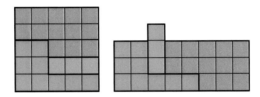

- In puzzle 2, the $5 \times 13$ rectangle can be covered by three puzzle pieces; one of the pieces protrudes with 1 unit from the $8 \times 8$ square.

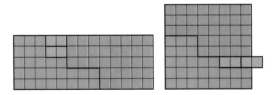

## 13.2   The Paradox of the Missing Square and the Right Angle Altitude Theorem of Euclid

Euclid's theorem of an altitude in a right triangle states that the square over an altitude is equal to the area of the rectangle whose sides are given by the segments of the hypotenuse. The converse of this theorem is correct, too, that is: only if the rectangle and the square have the same area, then the triangle is right-angled.

**Puzzle 4**

In the following figure a square is drawn above the altitude with the area $8 \cdot 8 = 64$ units and also a rectangle with the segments of the hypotenuse as sides; the rectangle's area is $5 \cdot 13 = 65$.

Solution: Since the two figures obviously do not have the same area, it follows that the triangle is not right-angled.

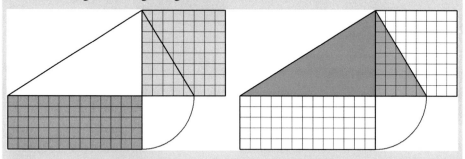

From the total sum of the measures of the interior angles we can calculate the real size of the angle:

$$\tan^{-1}\left(\tfrac{13}{8}\right) + \tan^{-1}\left(\tfrac{5}{8}\right) = \left[90^{\circ} - \tan^{-1}\left(\tfrac{8}{13}\right)\right] + \tan^{-1}\left(\tfrac{5}{8}\right)$$

$$\approx (90^{\circ} - 31.61^{\circ}) + 32.01^{\circ} = 58.39^{\circ} + 32.01^{\circ} = 90.40^{\circ}$$

By the way, the last figure contains the same figures as in the illustrations to puzzle 2 (see figure on the right): the light-blue triangle with the legs of 5 and 8 length units, the green triangle with the legs of 13 and 8 length units.

Analogously you can draw an almost right-angled triangle to illustrate puzzle 3.

**Puzzle 5**

The following figure again contains the "proof" that it applies: $3 \cdot 8 = 5^2$.

Solution: The apparently right angle actually has the following size:

$$\tan^{-1}\left(\tfrac{8}{5}\right) + \tan^{-1}\left(\tfrac{3}{5}\right) \approx 57.99^{\circ} + 30.96^{\circ} = 88.95^{\circ}.$$

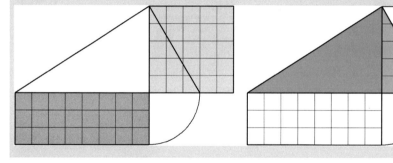

And the next picture contains a figure that applies to puzzle 1:

$$\tan^{-1}\left(\tfrac{21}{13}\right) + \tan^{-1}\left(\tfrac{8}{13}\right) \approx (90° - 31.76°) + 31.61° = 58.24° + 31.61° = 89.85°.$$

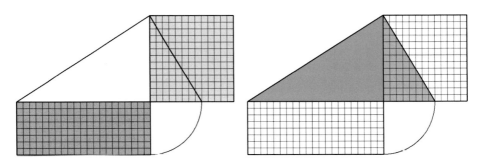

When we choose the side lengths along the sequence of Fibonacci numbers, the angle in the triangle is alternately smaller and larger than 90°, and it approaches more and more a right angle (see the following table).

| Rectangle | 3x8 | 5x13 | 8x21 | 13x34 | 21x55 | ... |
|---|---|---|---|---|---|---|
| Square | 5x5 | 8x8 | 13x13 | 21x21 | 34x34 | ... |
| Angle | 88.95° | 90.40° | 89.85° | 90.06° | 89.98° | ... |
| Deviation from 90° | 1.05° | 0.40° | 0.15° | 0.06° | 0.02° | ... |

The next figures show the almost right-angled triangles corresponding to the Fibonacci numbers 13, 21, 34 and to 21, 34, 55, for which the angle opposite to the longest side in the triangle approaches closer and closer to 90°.

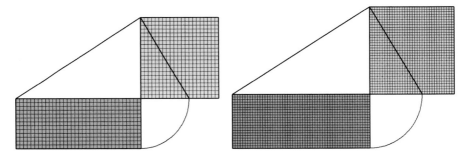

Usually one checks whether a triangle is right angled using the Pythagorean theorem. To do this, one would have to calculate the length of the legs $a$ and $b$ (see the following figure).

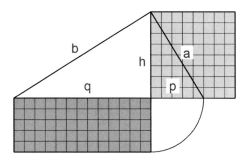

For the basic side $c$ of the triangle ("hypotenuse") we have in puzzle 4:
$c = p + q = 5 + 13 = 18$, so $c^2 = 18^2 = 324$.
For the "legs" it applies:
$a^2 = p^2 + h^2 = 5^2 + 8^2 = 89$ and
$b^2 = h^2 + q^2 = 8^2 + 13^2 = 233$, so $a^2 + b^2 = 322$.

The sum $a^2 + b^2 = 322$ is smaller by 2 than $c^2 = 324$.

So as $a^2 + b^2 < c^2$, follows after the generalization of the Pythagorean theorem, that the angular size of $\gamma$ must be greater than $90°$.

The angular size of $\gamma$ can be determined using the *law of cosines*:

$$\cos^{-1}(\gamma) = \cos^{-1}\left(\frac{a^2 + b^2 - c^2}{2ab}\right)$$

Here is $\cos^{-1}(\gamma) = \cos^{-1}\left(\frac{-2}{2 \cdot \sqrt{89} \cdot \sqrt{233}}\right) = \cos^{-1}\left(\frac{-1}{\sqrt{89} \cdot \sqrt{233}}\right) \approx 90.40°$ (see table above).

Correspondingly, it applies for the side lengths in puzzle 5:

$a^2 = 3^2 + 5^2 = 34;\ \ b^2 = 5^2 + 8^2 = 89;\ \ a^2 + b^2 = 123$
$c^2 = (3 + 8)^2 = 11^2 = 121$
$\cos^{-1}(\gamma) = \cos^{-1}\left(\frac{2}{2 \cdot \sqrt{34} \cdot \sqrt{89}}\right) = \cos^{-1}\left(\frac{-1}{\sqrt{34} \cdot \sqrt{89}}\right) \approx 88.96°$ (see table above)

Astonishing fact: $c^2$ differs from $a^2 + b^2$ by 2, too.

But there is also another remarkable feature:

$a^2$ and $b^2$ are on the one hand sums of squares of two consecutive Fibonacci numbers, but on the other hand they are themselves elements of the Fibonacci sequence (see the following table with the first 16 Fibonacci numbers).

| $n$ | 1 | 2 | 3 | 4 | 5 | 6 | 7 | 8 | 9 | 10 | 11 | 12 | 13 | 14 | 15 | 16 | ... |
|---|---|---|---|---|---|---|---|---|---|---|---|---|---|---|---|---|---|
| $f_n$ | 1 | 1 | 2 | 3 | 5 | 8 | 13 | 21 | 34 | 55 | 89 | 144 | 233 | 377 | 610 | 987 | ... |

It's valid:

$$f_1^2 + f_2^2 = 1^2 + 1^2 = 2 = f_3; f_2^2 + f_3^2 = 1^2 + 2^2 = 5 = f_5; f_3^2 + f_4^2 = 2^2 + 3^2 = 13 = f_7;$$

$$f_4^2 + f_5^2 = 3^2 + 5^2 = 34 = f_9; f_5^2 + f_6^2 = 5^2 + 8^2 = 89 = f_{11};$$
$$f_6^2 + f_7^2 = 8^2 + 13^2 = 233 = f_{13}; \ldots$$

In fact, there is a general rule:

---

**Formula**

For the squares of two consecutive numbers $f_n$ and $f_{n+1}$ of the Fibonacci sequence the following is true:

$$f_n^2 + f_{n+1}^2 = f_{2n+1} \tag{13.2}$$

◀

---

You can use this relationship to determine large Fibonacci numbers without calculating all the intermediates.

---

**Example: Determination of $f_{31}$**
The last entries in the table above are $f_{15}$ and $f_{16}$. With this you can calculate $f_{31}$:

$$f_{31} = f_{15}^2 + f_{16}^2 = 610^2 + 987^2 = 1346269$$

---

For the difference between $a^2 + b^2$ and $c^2$ you have:
$a^2 = p^2 + h^2 = f_{n-1}^2 + f_n^2 (= f_{2n-1})$ and
$b^2 = h^2 + q^2 = f_n^2 + f_{n+1}^2 (= f_{2n+1})$ as well as
$c^2 = (p+q)^2 = (f_{n-1} + f_{n+1})^2 = f_{n-1}^2 + 2 \cdot f_{n-1} \cdot f_{n+1} + f_{n+1}^2$, that is
$a^2 + b^2 - c^2 = f_{n-1}^2 + f_n^2 + f_n^2 + f_{n+1}^2 - f_{n-1}^2 - 2 \cdot f_{n-1} \cdot f_{n+1} - f_{n+1}^2$
$$= 2 \cdot (f_n^2 - f_{n-1} \cdot f_{n+1}) = 2 \cdot (-1)^{n+1} = \pm 2$$

according to Cassini's identity $f_{n-1} \cdot f_{n+1} - f_n^2 = (-1)^n$ (see formula 13.1).
In fact, the difference between $a^2 + b^2$ and $c^2$ is always 2.
For the formula to calculate the largest angle of the almost right-angled triangle, therefore, you have:

$$\cos^{-1}(\gamma) = \cos^{-1}\left(\frac{a^2 + b^2 - c^2}{2ab}\right) = \cos^{-1}\left(\frac{(-1)^{n+1}}{\sqrt{f_{2n-1} \cdot f_{2n+1}}}\right)$$

$$= \cos^{-1}\left(\frac{(-1)^{n+1}}{\sqrt{f_{2n}^2 + (-1)^{2n}}}\right) = \cos^{-1}\left(\frac{(-1)^{n+1}}{\sqrt{f_{2n}^2 + 1}}\right)$$

This formula shows that the angle is alternately larger and smaller than $90°$ and with increasing $n$ it converges to $90° = \cos^{-1}(0)$.

## 13.3   The Missing Square and Other Methods of Euclid

The application of the law of altitudes is one of the geometrical methods to convert a square into a rectangle of equal area or a rectangle into a square of equal area. In the *Elements* of Euclid, one also finds other methods.

### 13.3.1  Application of Euclid's Theorem

The square on a side of a right-angled triangle next to the right angle has the same area as the rectangle formed by the hypotenuse and the segment of the hypotenuse cut off by the altitude from the right angle that corresponds to that side (Euclid's theorem). Conversely, one can conclude from the fact that the areas of rectangle and square are equal that the triangle is right-angled.

While in the first of the following two figures (Fibonacci numbers 3, 5, 8) one might still be able to see that there is a "kink" in the line consisting of the other side and the square side, this is hardly noticeable in the figure on the right (Fibonacci numbers 8, 13, 21).

The triangles shown are not right-angled, because the right part of the triangle is not similar to the total triangle: In the figure on the left, this can be seen from the ratios of the side lengths of the legs of the triangle and hypotenuse 3:5 and 5:8 respectively; in the figure on the right, these are the ratios 8:13 and 13:21 respectively.

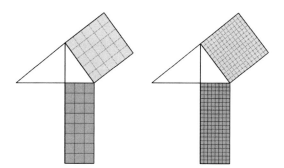

**Suggestions for Reflection and for Investigations**
**A 13.4:** Is it possible to draw the green or light-blue triangle from the figures of puzzle 1 or puzzle 3 into the last graphics?

### 13.3.2  Application of Areas

As is well known, the product of two sums

$$(a + b) \cdot (c + d) = a \cdot c + a \cdot d + b \cdot c + b \cdot d$$

can be displayed geometrically using rectangles (see the following figure).

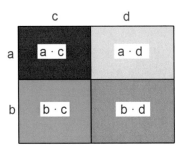

The area of the rectangle with the sides $a$ and $c$ is equal to the area of the rectangle with the sides $b$ and $d$ if and only if the common vertex of the two rectangles is lying on the diagonal $PR$. Since then, because of the similarity of the triangles lying below and above the diagonal, the equation of ratios $b{:}c = a{:}d$, that is $b \cdot d = a \cdot c$ holds.

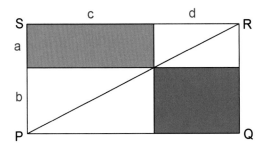

This property can be used constructively to transform a rectangle into a square of equal area (and vice versa). The procedure is called an *application of areas*.

In the first of the following two graphics, the green and the light-blue rectangle apparently have a common diagonal, but in fact this is not right. But if the side lengths are numbers of the first elements of the Fibonacci sequence you have a real chance to spot the "kink" between the two segments (see figure on the right).

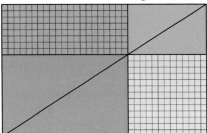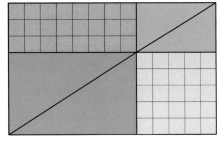

## 13.4   Other Properties in Connection with Fibonacci Numbers

Instead of considering three consecutive elements $f_{n-1}, f_n, f_{n+1}$ of the Fibonacci sequence and using the relationship $f_{n-1} \cdot f_{n+1} = f_n^2 \pm 1$ (see formula 13.1) for the paradox, one can also select *four* consecutive numbers of the sequence and visualize the property that the quotients are approximately equal.

If you consider the first and the last quotient of $\frac{f_n}{f_{n-1}} \approx \frac{f_{n+1}}{f_n} \approx \frac{f_{n+2}}{f_{n+1}}$ then you get $f_n \cdot f_{n+1} \approx f_{n-1} \cdot f_{n+2}$.

---

**Example**

For the Fibonacci numbers 3, 5, 8, 13 it follows:
$$39 = 3 \cdot 13 \approx 5 \cdot 8 = 40$$
For 5, 8, 13, 21 you have:
$$105 = 5 \cdot 21 \approx 8 \cdot 13 = 104$$
And for 8, 13, 21, 34:
$$272 = 8 \cdot 34 \approx 13 \cdot 21 = 273$$

The product of the two inner elements of the Fibonacci sequence is greater or smaller by 1 than the product of the two outer elements.

---

If one replaces $f_{n+2}$ and $f_{n+1}$ according to definition and replaces the product $f_{n-1} \cdot f_{n+1}$ throughout by $f_n^2 + (-1)^n$ according to Cassini's identity, then, in fact, the following results:

$$\begin{aligned}
f_{n-1} \cdot f_{n+2} - f_n \cdot f_{n+1} &= f_{n-1} \cdot (f_n + f_{n+1}) - f_n \cdot f_{n+1} \\
&= f_{n-1} \cdot f_n + f_{n-1} \cdot f_{n+1} - f_n \cdot (f_{n-1} + f_n) \\
&= f_{n-1} \cdot f_n + \left(f_n^2 + (-1)^n\right) - f_n \cdot f_{n-1} - f_n^2 = (-1)^n
\end{aligned}$$

---

**Formula**

For four consecutive numbers $f_{n-1}, f_n, f_{n+1}, f_{n+2}$ of the Fibonacci sequence it follows:
$$f_{n-1} \cdot f_{n+2} - f_n \cdot f_{n+1} = (-1)^n \tag{13.3}$$

◄

Formula (13.3) is a special case of a general formula proved by the French mathematician Philbert Maurice d'Ocagne (1862–1938), which is known in literature as *Identity of d'Ocagne*.

So this relationship between four consecutive Fibonacci numbers can also be used to make a square unit disappear:

If you draw two right-angled triangles with $f_{n-1}$ and $f_{n+1}$ as lengths of the perpendicular sides or with $f_n$ and $f_{n+2}$, then – if $n$ is sufficiently large – it looks as if the hypotenuses in these triangles have the same slope (see the following table and the two illustrations).

| Slope | $\frac{3}{8} = 0.375$ | $\frac{5}{13} = 0.384...$ | $\frac{8}{21} = 0.380...$ | $\frac{13}{34} = 0.382...$ | $\frac{21}{55} = 0.381...$ | $\frac{34}{89} = 0.382...$ |
|-------|------------|--------------|--------------|----------------|----------------|----------------|
| Angle | 20.56° | 21.04° | 20.85° | 20.92° | 20.898° | 20.908° |

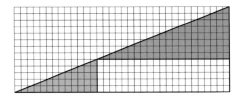 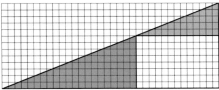

**Suggestions for Reflection and for Investigations**

**A 13.5:** Determine the limit of the sequence defined by $\frac{f_n}{f_{n+2}}$.

**A 13.6:** Analyze "Curry's triangle paradox" in the figures at the beginning of the chapter.

**A 13.7:** Analyze the following figures.

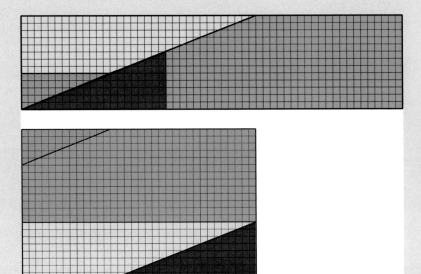

## 13.5   Arrangement by Sam Loyd

In Sam Loyd's arrangement, the triangles considered in Sect. 13.4 are put together with
trapezoidal puzzle pieces to form a rectangle or a square.

Loyd's arrangement is

- for the Fibonacci numbers 5, 8, 13: $8^2 = 64 < 65 = 5 \cdot 13$

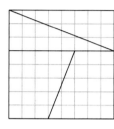 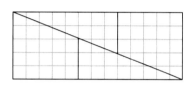

- for the Fibonacci numbers 8, 13, 21: $13^2 = 169 > 168 = 8 \cdot 21$

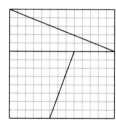 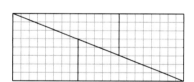

- for the Fibonacci numbers 13, 21, 34: $21^2 = 441 < 442 = 13 \cdot 34$

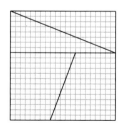 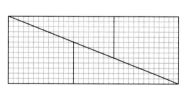

If you analyze the graphics closely, you can see – at least at the first figure – that you can
use the grid in the background to draw the squares on the left, but not the rectangles on
the right.

## 13.6   Other Appropriate Triples of Numbers

The paradox of the missing square can be observed with numbers other than those of the Fibonacci sequence. In principle, it is sufficient to find three numbers $a$, $b$, and $c$ so that the rectangle with the side lengths $a$ and $b$ is approximately as large as the square with the side length $c$. You can do this by reviewing the list of square numbers $c^2$ and checking whether $c^2 - 1$ or $c^2 + 1$ can be factorized as $a \cdot b$.

Then the gradient angle in the right-angled triangles with the perpendicular sides $a$ and $c$ or rather $c$ and $b$ is examined. In Fig. 13.1 only those rectangles are listed whose shorter side length is at least 2 units. Because of the binomial formula, for $c^2 - 1$ the decomposition $(c - 1) \cdot (c + 1)$ applies always.

Then you can draw right triangles with $\alpha = \tan^{-1}\left(\frac{c}{a}\right)$ and $\beta = \tan^{-1}\left(\frac{b}{c}\right)$.

In the following graphics the example $a = 5$, $b = 16$, $c = 9$ is illustrated (compare the table in Fig. 13.1).

| $c$ | $c^2$ | $c^2 - 1$ | $a$ | $b$ | $\alpha$ | $\beta$ | $c^2 + 1$ | $a$ | $b$ | $\alpha$ | $\beta$ |
|---|---|---|---|---|---|---|---|---|---|---|---|
| 3 | 9 | 8 | 2 | 4 | 33,69° | 36,87° | 10 | 2 | 5 | 33,69° | 30,96° |
| 4 | 16 | 15 | 3 | 5 | 36,87° | 38,66° | 17 | | | | |
| 5 | 25 | 24 | 4 | 6 | 38,66° | 39,81° | 26 | 2 | 13 | 21,80° | 21,04° |
| | | | 3 | 8 | 30,96° | 32,01° | | | | | |
| | | | 2 | 12 | 21,80° | 22,62° | | | | | |
| 6 | 36 | 35 | 5 | 7 | 39,81° | 40,60° | 37 | | | | |
| 7 | 49 | 48 | 6 | 8 | 40,60° | 41,19° | 50 | 5 | 10 | 35,54° | 34,99° |
| | | | 4 | 12 | 29,74° | 30,26° | | 2 | 25 | 15,95° | 15,64° |
| | | | 3 | 16 | 23,20° | 23,63° | | | | | |
| | | | 2 | 24 | 15,95° | 16,26° | | | | | |
| 8 | 64 | 63 | 7 | 9 | 41,19° | 41,63° | 65 | 5 | 13 | 32,01° | 31,61° |
| | | | 3 | 21 | 20,56° | 20,85° | | | | | |
| 9 | 81 | 80 | 8 | 10 | 41,63° | 41,99° | 82 | 2 | 41 | 12,38° | 12,41° |
| | | | 5 | 16 | 29,05° | 29,36° | | | | | |
| | | | 4 | 20 | 23,96° | 24,23° | | | | | |
| | | | 2 | 40 | 12,53° | 12,68° | | | | | |
| 10 | 100 | 99 | 9 | 11 | 41,99° | 42,27° | 101 | | | | |
| | | | 3 | 33 | 16,70° | 16,86° | | | | | |

**Fig. 13.1** Suitable triples of numbers to demonstrate the missing square's paradox

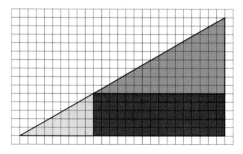 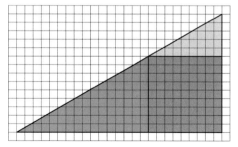

## 13.7   The Missing Square and the Pythagorean Theorem

In the previous examples, geometric figures such as triangles, rectangles, and squares were considered, in which differences of 1 unit (or even more) occurred when parts of the figures were moved. In all examples, the "missing square" could be exposed as an illusion due to slightly different angles.

In connection with the well-known ffigure of Pythagoras's theorem (more on this in Chap. 17), a difference of 1 unit can also occur, namely if the angle $\gamma$, which is opposite to the longest side in the triangle, deviates slightly from $90°$.

By systematic trial and error one finds the ten triples of side lengths of the $a$, $b$ and the "hypotenuse" $c$ listed in the table of Fig. 13.2 which lengths are not greater than 20 length units and for which $a^2 + b^2$ differs by exactly 1 square unit from $c^2$.

In each example the deviations from the right angle are so small that one only realizes that the triangle sides are not extensions of the square sides if one has counted the boxes.

Among the examples of Fig. 13.2 there are also two "symmetrical" examples with isosceles triangles. These triples $(5, 5, 7)$ and $(12, 12, 17)$ are shown in the following figures with the chessboard-like structure.

**Fig. 13.2** Almost-Pythagoras-figures with difference 1 square unit (ordered by the length of the "hypotenuse")

| $a$ | $b$ | $c$ | $a^2 + b^2$ | $c^2$ | $\gamma$ |
|---|---|---|---|---|---|
| 5 | 5 | 7 | 50 | 49 | 88,85° |
| 4 | 7 | 8 | 65 | 64 | 88,98° |
| 4 | 8 | 9 | 80 | 81 | 90,90° |
| 8 | 9 | 12 | 145 | 144 | 89,60° |
| 7 | 11 | 13 | 170 | 169 | 89,63° |
| 11 | 13 | 17 | 290 | 289 | 89,80° |
| 12 | 12 | 17 | 288 | 289 | 90,20° |
| 6 | 17 | 18 | 325 | 324 | 89,72° |
| 10 | 15 | 18 | 325 | 324 | 89,81° |
| 6 | 18 | 19 | 360 | 361 | 90,27° |

The fact that the number of unit squares differs by 1 unit is illustrated by coloring subareas with different colors.

## 13.8   References to Further Literature

On **Wikipedia** you can find further information and literature on the keywords in English (German, French):

- Missing square puzzle (Fehlendes Quadrat Rätsel, Paradoxe du carré manquant),
- Samuel Loyd (Sam Loyd, Sam Loyd)

More informations can be found at **Wolfram Mathworld** under the keywords:

- Curry Triangle, Dissection Fallacy, Triangle Dissection Paradox

Alexander Bogomolny's website is also recommended

- https://www.cut-the-knot.org/Curriculum/Fallacies/CurryParadox.shtml

as well as Hans Walser's "Miniaturen" (https://www.walser-h-m.ch/hans/Miniaturen/) on this subject:

- Verschwundenes Quadrat (missing square), Fastpythagoreische Dreiecke (almost-Pythagorean-triangles)

# Dissection of Rectangles into Squares of Different Sizes

# 14

*Mathematics, rightly viewed, possesses not only truth, but supreme beauty – a beauty cold and austere, like that of sculpture.*

*(Bertrand Russell, British philosopher and Mathematician, 1872–1970)*

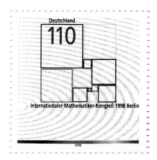

For over 100 years mathematicians have been studying the question, whether it is possible to dissect a square *into squares of different sizes*; such squares are called *simple perfect squared squares*. (Note the difference to the question in Chap. 3, where the aim is to dissect a rectangle into *largest possible* squares.

The stamp issued by Deutsche Post on the occasion of the International Congress of Mathematicians (Internationaler Mathematikerkongress - ICM) in Berlin in 1998 (see above) refers to this question.

In 1925, the Polish mathematician Zbigniew Moroń (1904–1971) was the first to succeed in the tessellating of *rectangles* using squares of different sizes *(simple perfect squared rectangles)*. Fourteen years later, it was the German mathematician Roland P. Sprague (1894–1967) who discovered the first "squaring the square" – covering a square

© Springer-Verlag GmbH Germany, part of Springer Nature 2021
H. K. Strick, *Mathematics is Beautiful*, https://doi.org/10.1007/978-3-662-62689-4_14

by 55 squares of different sizes. Finally, in 1962, the Dutch mathematician Adrianus Johannes Wilhelmus Duijvestijn (1927–1998) proved that you need at least 21 squares of different sizes to tessellate a square. And it was he who finally found such a square with the help of a computer – but only 16 years later.

In this chapter, we will concentrate on the problem to analyze rectangles and squares which have been tessellated; for further possible investigations, please refer to the bibliography.

*By the way:* For coloring the split rectangles and squares you need at most four colors – this so-called *four color theorem* was proven in 1976. This theme can also be found on the ICM stamp (and in the background: the decimal development of the number $\pi$).

---

## 14.1  Rectangles which can be Dissected into Nine or Ten Squares of Different Sizes

To tessellate a rectangle only with squares of different sizes, you need at least nine squares. The two diagrams in the Fig. 14.1 show the only rectangles that can be tessellated with *nine* squares of different sizes. On the left we have a rectangle that is covered by squares with the side lengths 1, 4, 7, 8, 9, 10, 14, 15, 18 and on the right a rectangle that is covered by squares with the side lengths 2, 5, 7, 9, 16, 25, 28, 33, 36.

The next two graphics in Fig. 14.2 show two of the six rectangles that can be tessellated with *ten* squares of different sizes. Here we have squares with the side lengths 2, 3, 8, 11, 13, 15, 17, 25, 27, 30 (Fig. 14.2a) and 3, 5, 6, 11, 17, 19, 22, 23, 24, 25 (Fig. 14.2b).

Which of the squares in the figures has which side length can be found out by simple considerations.

a 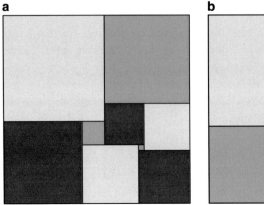      b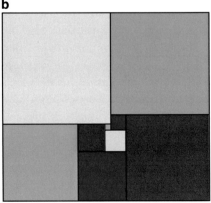

**Fig. 14.1  a, b** Tessellation of rectangles with nine squares of different sizes

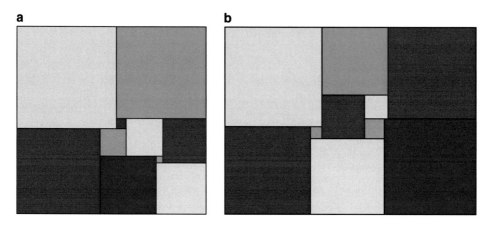

**Fig. 14.2  a, b** Tessellation of rectangles with ten squares of different sizes

**Example: How to attribute the specified side lengths to the right squares**

With the information about the side lengths 1, 4, 7, 8, 9, 10, 14, 15, 18 of the squares in Fig. 14.1a the following results:

The smallest square with the side length 1 (bottom right, green) is easy to identify, also the squares with the side lengths 4 (center, green), 7 (center, red) and 8 (center right, yellow).

Furthermore the following applies to the adjacent squares: 9 (bottom right, red), 10 (bottom, yellow). By comparing with the side length of adjacent squares we find the squares with the side lengths 14 (bottom left, red) = 4 (green) + 10 (yellow), 15 (top right, green) = 7 (red) + 8 (yellow) and finally the square with the side length 18 (top left, yellow) = 14 (red) + 4 (green).

In order to check the assigned side lengths, the sum of the side lengths should be calculated at various positions horizontally and vertically, thus also determining the total size of the rectangle.

Vertical cuts through the figure (noted from top to bottom):

- 18 (yellow) + 14 (red) = 32;
- 18 (yellow) + 4 (green) + 10 (yellow) = 32;
- 15 (green) + 7 (red) + 10 (yellow) = 32;
- 15 (green) + 8 (yellow) + 9 (red) = 32.

Horizontal cuts through the figure (noted from left to right):

- 18 (yellow) + 15 (red) = 33;
- 18 (yellow) + 7 (red) + 8 (yellow) = 33;
- 14 (red) + 4 (green) + 7 (red) + 8 (yellow) = 33;
- 14 (red) + 10 (yellow) + 9 (red) = 33.

Therefore the rectangle in Fig. 14.1a has the width 33 and the height 32.

This is the tessellation of the rectangle which was found by Zbigniew Moroń in 1925. The side lengths of the squares can also be seen in the following figure – except for the square with side length 1.

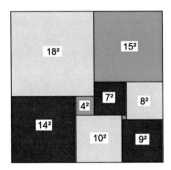

**Suggestions for reflection and for investigations**
**A 14.1:** Show that the rectangle in Fig. 14.1b has the dimensions $69 \times 61$, the rectangle in Fig. 14.1a the dimensions $57 \times 55$ and the rectangle in Fig. 14.2b the dimensions $65 \times 47$.

## 14.2   Determining the Side Lengths for a given Tessellation

Checking the side length of particular squares, when all side lengths are known, has turned out to be a simple exercise. Is it also possible to find out the side lengths if only a picture of the tessellation is available, but no side lengths are known?

This is the task we want to face now. Using the example of the tessellation on the ICM stamp (see above, Fig. 14.3a), the following explains a procedure that is so elementary that it can be applied for introducing the use of expressions in arithmetic.

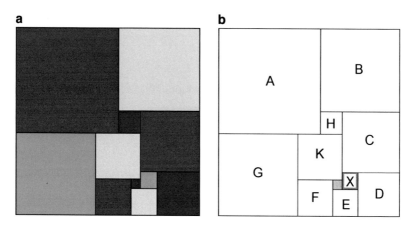

**Fig. 14.3  a, b** Rectangle from the ICM stamp – approach for determining side lengths

The smallest square (light gray) is first getting assigned to the side length 1, the side length of the second smallest square (dark gray) is designated with a variable ($x$). The remaining squares are labeled with the variables $A$ to $K$ (see Fig. 14.3b).

Now we have to determine the side lengths of the remaining squares using 1 and $x$.

The first steps are easy; for the side length of the adjacent square $E$ it applies:

1. $e = 1 + x$ (in words: the side length of the square $E$ is by 1 unit larger than the side length of the dark gray marked square).
   Further we have
2. $f = e + 1$
   Using the representation from (1) we get $f = (1 + x) + 1$, that is $(2)'f = 2 + x$.
   and onwards
3. $k = f + 1 = (2 + x) + 1 = 3 + x$ and
4. $g = k + f = (3 + x) + (2 + x) = 5 + 2 \cdot x$ (in words: the side length of the square $G$ is 5 units larger than twice the side length of the second smallest square).
   By convention, the multiplication sign is omitted for $2 \cdot x$.
5. $d = e + x = (1 + x) + x = 1 + x + x = 1 + 2x$
6. $c = d + x = (1 + 2x) + x = 1 + 2x + x = 1 + 3x$
   When determining the side length of square $H$, we see: $h$ can only be determined by comparing the side lengths of several adjacent squares.
   As $c + d = h + k + f$ we can conclude
7. $h = (c + d) - (k + f) = [(1 + 3x) + (1 + 2x)] - [(3 + x) + (2 + x)]$
$$= [1 + 3x + 1 + 2x] - [3 + x + 2 + x]$$
$$= [2 + 5x] - [5 + 2x] = 2 + 5x - 5 - 2x = 3x - 3$$

And further:

8. $b = c + h = (1 + 3x) + (3x - 3) = 6x - 2$
9. $a = b + h = (6x - 2) + (3x - 3) = 9x - 5$

Finally we compare the side lengths of the total figure at various positions:

The height of the rectangle on the left side is:

$$a + g = (9x - 5) + (5 + 2x) = 9x - 5 + 5 + 2x = 11x$$

The height of the rectangle on the right side is

$$b + c + d = (6x - 2) + (1 + 3x) + (1 + 2x) = 11x$$

In both cases, we get the same term – namely that the height of the rectangle is 11 times the side length of the second smallest square.

The width of the rectangle at the top side is

$$a + b = (9x - 5) + (6x - 2)$$
$$= 9x - 5 + 6x - 2$$
$$= 15x - 7$$

and at the bottom

$$g + f + e + d = (5 + 2x) + (2 + x) + (1 + x) + (1 + 2x)$$
$$= 5 + 2x + 2 + x + 1 + x + 1 + 2x$$
$$= 6x + 9$$

The fact that we get different expressions for the width of the rectangle means that the two terms must be equal. So we have:

$$15x - 7 = 6x + 9$$

And further by transformations the following results

$$15x = 6x + 16$$

and from this:

$$9x = 16$$

This means that the side length of the second smallest square is exactly 16/9 times the side length of the smallest square. If we had not chosen 1 length unit as side length at the beginning, but nine times it, that is, 9 length units, then *integral* side lengths would have occurred in all squares of the figure.

The side lengths of the individual squares are: 9, 16, 21, 25, 34, 41, 43, 57, 77, 78, and 99. The figure depicted on the stamp is, therefore, actually a rectangle with the side lengths $176 \times 177$.

**Fig. 14.4** Calculation of the side lengths of the squares on the ICM stamp

| Term for the side length | Side length | 9 times the side length |
|---|---|---|
| $x$ | $x = 16/9$ | $9x = 16$ |
| $a = 9x - 5$ | $a = 99/9$ | $9a = 99$ |
| $b = 6x - 2$ | $b = 78/9$ | $9b = 78$ |
| $c = 1 + 3x$ | $c = 57/9$ | $9c = 57$ |
| $d = 1 + 2x$ | $d = 41/9$ | $9d = 41$ |
| $e = 1 + x$ | $e = 25/9$ | $9e = 25$ |
| $f = 2 + x$ | $f = 34/9$ | $9f = 34$ |
| $g = 5 + 2x$ | $g = 77/9$ | $9g = 77$ |
| $h = 3x - 3$ | $h = 21/9$ | $9h = 21$ |
| $k = 3 + x$ | $k = 43/9$ | $9k = 43$ |
| Total height: $11x$ | $176/9$ | $176$ |
| Total width $15x - 7 = 6x - 9$ | $177/9$ | $177$ |

In the table of Fig. 14.4, the steps for the calculation and the final results are shown again for the sake of a better overview.

The method used here for expressing the side lengths of squares using the side lengths of adjacent squares can also be used in other examples. Depending on the position of the two adjacent squares to start with, it happens that different expressions arise for the widths and heights of the total rectangles (as we have seen above). It is also possible to find different expressions just for one of the squares, so that the variable $x$ can be determined already from this condition.

Actually, this is a problem that requires two variables to solve; but with the "trick" of assuming that the smallest square is a unit square, the number of variables is reduced to 1. As in the example, the solutions can finally be multiplied by a suitable factor to find the smallest *integral* solution among the infinite number of (mutually proportional) solutions.

**Suggestions for reflection and for investigations**
**A 14.2:** Determine relationships between the side lengths of the rectangle shown in Fig. 14.1a. (The squares in the following figure are already labeled in order to compare your approach with other solutions.)

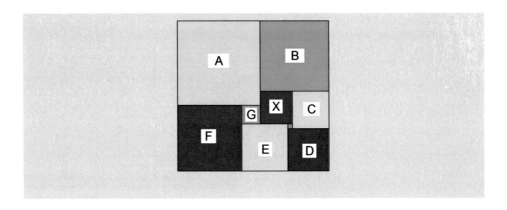

## 14.3   Introduction of the Bouwkamp notation to describe a tessellation

Christoffel Jacob Bouwkamp (1915–2003), the doctoral supervisor of Duijvestijn, introduced a notation for the rectangular tessellations that can be used to describe how to draw a rectangle that is dissected by squares.

**Example: Description of the tessellation of the 69 × 61 rectangle from Fig. 14.1b**
In Bouwkamp's notation, the rectangular tessellation is described by the term (36,33)(5,28)(25,9,2)(7)(16).

The side lengths of the squares (and the total rectangle) can be read from the following figure (the small green square in the center has the side length 2).

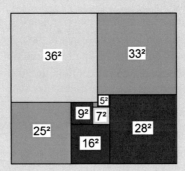

You "translate" the Bouwkamp notation by moving through the figure from left to right and from top to bottom:

(36,33) – you start to draw a square with the side length 36 and attach a square of side length 33 to the right of it so that the two top sides form a continuous line.

The square of the side length 33 does not reach as far down as the square of the side length 36. Seen from top to bottom, now squares will follow, which connect to the square of the side length 33.

(5,28) – The next two noted squares with the side lengths 5 and 28 attach exactly to the left and above to the square with the side length 33.

(25,9,2) – The next squares with side lengths 25, 9, and 2 (green, without a label) are then drawn below the square of the side length 36.

(7) – Of the squares drawn so far, the squares with the side lengths 2 and 5 finish below "at the same height". Into the gap between the square with the side length 28, which is already drawn, a square of the side length 7 fits exactly.

(16) – The squares with side lengths 9 or 7 have a common lower edge so that finally the square with side length 16 can be inserted into the figure.

Besides the two rectangles in Fig. 14.2 there are four more rectangles that can be dissected into ten squares of different sizes (see Fig. 14.5).

And there are 67 rectangles which can be dissected into 12 squares of different sizes, 213 rectangles which can be covered by 13 squares of different sizes, 744 rectangles each dissected by 14 squares of different sizes, 2,609 rectangles made up of 15 squares of different sizes ...

**Fig. 14.5  a–d** More examples of rectangles, which can be dissected into ten squares of different sizes

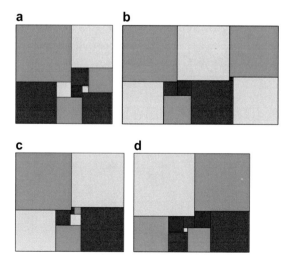

**Suggestions for reflection and for investigations**

**A 14.3:** Determine the Bouwkamp notation for the tessellation of the

1. $33 \times 32$ rectangle from Fig. 14.3,
2. $57 \times 55$ rectangle from Fig. 14.6a,
3. $65 \times 47$ rectangle from Fig. 14.6b.

**A 14.4:** Determine the side lengths of the rectangle and of the ten squares for the rectangles in Fig. 14.5a–d. (The squares have already been labeled).

Note the description of the tessellation using the Bouwkamp notation.

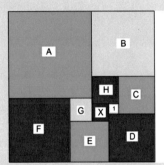 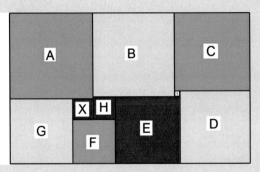

**A 14.5:** There are also 22 rectangles that can be tessellated by eleven squares of different sizes (one of the 22 rectangles is the rectangle of the ICM stamp).

Draw the following rectangles:

1. $97 \times 96$ rectangle: $(56,41)(17,24)(40,14,2)(12,7)(31)(26)$
2. $98 \times 86$ rectangle: $(51,47)(8,39)(35,11,5)(1,7)(6)(24)$
3. $98 \times 95$ rectangle: $(50,48)(7,19,22)(45,5)(12)(28,3)(25)$
4. $112 \times 81$ rectangles: $(43,29,40)(19,10)(9,1)(41)(38,5)(33)$

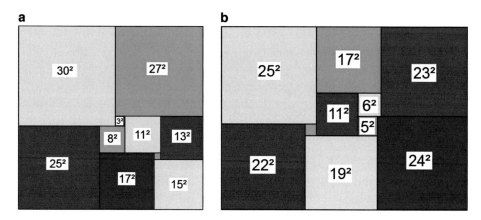

**Fig. 14.6  a, b** On the left the tessellation of the $57 \times 55$ rectangle into squares of different sizes (the small green square at the bottom right has the side length 2), on the right, the $65 \times 47$ rectangle is dissected into squares (the small green square in the center has a side length 3)

## 14.4    Squares, which can be Dissected into Squares of Different Sizes

After R. P. Sprague had discovered the first "squared square" in 1939, consisting of 55 squares of different sizes, a "competition" started to find the least number of squares required. As described above, A. J. W. Duijvestijn finally found in 1978 a square which can be tessellated by only 21 squares of different sizes – after he had been able to prove years earlier that this number cannot be reduced even further.

This square has the format $112 \times 112$ and the following Bouwkamp notation: $(50,35,27)(8,19)(15,17,11)$ $(6,24)(29,25,9,2)(7,18)(16)(42)(4,37)(33)$ (see Fig. 14.7a).

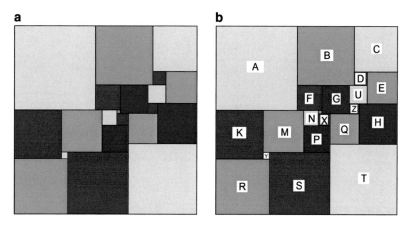

**Fig. 14.7  a, b** The smallest square that can be tessellated using 21 squares of different sizes

Is it also possible to determine the side lengths of the squares by a simple algebraic method as we have seen before?

If you select a square to which you assign the side length 1, and an adjacent square with the side length $x$ then it is not possible to represent the side lengths of the other squares using only this variable. More variables are needed; the relations between the variables result in a linear system of equations that can be solved. With the help of the solutions, the remaining side lengths can then be determined.

---

**Example: Determination of the side lengths of the squares using Fig. 14.7b**

For example, if you choose the side length 1 for the smallest square (in the center, green) and the side length $x$ for the adjacent (red) square, you will find the following relationships:

$$n = x + 1 \qquad p = n + x = 2x + 1 \qquad m = n + p = 3x + 2$$

From the next step on, another variable $(y)$ is needed:

$$k = m + y = 3x + y + 2 \qquad r = k + y = 3x + 2y + 2 \qquad s = r + y = 3x + 3y + 2$$

You can use the variables $u$ and $z$ to describe the side lengths of the remaining squares:

$$g = u + z$$
$$f = g - 1 = u + z - 1 \qquad\qquad q = 1 + g + z - x = -x + u + 2z + 1$$
$$h = q + z = -x + u + 3z + 1 \qquad t = q + h = -2x + 2u + 5z + 2$$
$$e = h + z - u = -x + 4z + 1 \qquad d = e - u = -x - u + 4z + 1$$
$$c = d + e = -2x - u + 8z + 2 \qquad b = c + d = -3x - 2u + 12z + 3$$

On the other hand:

$$b = f + g + u - d = (u + z - 1) + (u + z) + u - (-x - u + 4z + 1)$$
$$= x + 4u - 2z - 2$$

From these two representations for $b$ therefore results

$$-3x - 2u + 12z + 3 = x + 4u - 2z - 2, \text{ that is}$$
$$\mathbf{4x + 6u - 14z = 5}$$

Further follows:

$$a = b + f = (-3x - 2u + 12z + 3) + (u + z - 1) = -3x - u + 13z + 2$$

On the other hand:

$$a = k + m + n + 1 - f = (3x + y + 2) + (3x + 2) + (x + 1) + 1 - (u + z - 1)$$
$$= 7x + y - u - z + 7$$

From both representations for $a$ then results

$$-3x - u + 13z + 2 = 7x + y - u - z + 7, \text{ that is}$$

$$\mathbf{10x + y - 14z = -5}$$

If you also note that the side lengths of the total figure must be the same on the left and on the right and also at the top and the bottom, then follows:

$$a + k + r = c + e + h + t, \text{ that is}$$

$$(7x + y - u - z + 7) + (3x + y + 2) + (3x + 2y + 2)$$
$$= (-2x - u + 8z + 2) + (-x + 4z + 1) + (-x + u + 3z + 1) + (-2x + 2u + 5z + 2), \text{ that is}$$

$$13x + 4y - u - z + 11 = -6x + 2u + 20z + 6 \text{ and hereby}$$

and

$$\mathbf{19x + 4y - 3u - 21z = -5}$$

$$a + b + c = r + s + t, \text{ that is}$$
$$(-3x - u + 13z + 2) + (-3x - 2u + 12z + 3) + (-2x - u + 8z + 2)$$
$$= (3x + 2y + 2) + (3x + 3y + 2) + (-2x + 2u + 5z + 2), \text{ that is}$$

$$-8x - 4u + 33z + 7 = 4x + 5y + 2u + 5z + 6 \text{ and with this:}$$

$$\mathbf{12x + 5y + 6u - 28z = 1}$$

The linear system of equations with four equations and four variables

$$4x + 0y + 6u - 14z = 5$$
$$10x + 1y + 0u - 14z = -5$$
$$19x + 4y - 3u - 21z = -5$$
$$12x + 5y + 6u - 28z = 1$$

has the solution.

   $x = 3.5; \; y = 2; \; u = 5.5 \text{ and } z = 3.$

   In order to obtain integral side lengths, one would have had to start from a smallest square of the side length 2, and then we get to the solution
$x = 7; \; y = 4; \; u = 11 \text{ and } z = 6.$

*Hint:* This is only an example of how to find the side lengths of the squares; other approaches are also possible.

   As we mentioned before the square just examined is composed of 21 squares of different sizes, and this is the smallest possible number of squares (also called "the smallest order"). Further examples:

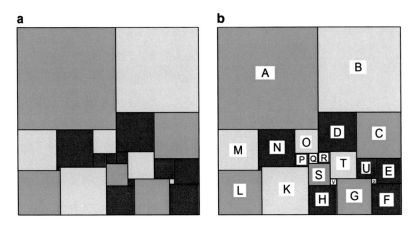

**Fig. 14.8  a, b** One of the two $110 \times 110$ squares, which can be tessellated using 22 squares of different sizes

- 8 square dissections of order 22 (an example is given in **A 14.6**),
- 12 square dissections of order 23,
- 26 square dissections of order 24,
- 160 square dissections of order 25,
- 441 square dissections of order 26,
- 1,152 square dissections of order 27,
- … (see references).

**Suggestions for reflection and for investigations**
**A 14.6:** Determine the side lengths of the square from Fig. 14.8 using a suitable system of equations.

## 14.5   Connection with Electrical Circuits

A systematic method for finding rectangular dissections was developed in the late 1930s by Rowland Leonard Brooks, Cedric Austin Bardell Smith, Arthur Harold Stone, and William Thomas Tutte, four students at Trinity College Cambridge.

The decisive factor was the idea of presenting the tessellations in a different way: horizontal lines are replaced by nodes (vertices) and these are connected with lines (or arrows).

In Fig. 14.9 this is shown by the example of the rectangle from Fig. 14.6a

**Fig. 14.9** Interpretation of the dissection of a rectangle as a network of electrical currents

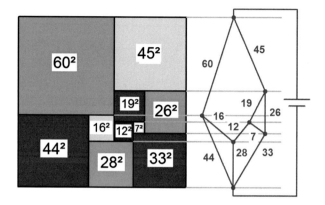

Now the resulted diagram can be interpreted as a network of electrical currents: The numbers on the lines indicate the respective intensity of the current flowing from top to bottom.

- The sum of the currents flowing to a node is equal to the sum of the currents flowing away from the node.

*Example of the top-right node:* 45 toward the node, $19 + 26$ away from the node.

- Looking at current loops within the network, the sum of the currents within a loop is zero if the direction is taken into account.

*Example for the top loop:* $60 + 16 - 12 - 19 - 45 = 0$.

These two statements correspond exactly to Kirchhoff's laws for electric circuits!

With the help of these two principles, it was then possible to systematically identify all possible examples using appropriate computer programs such as those developed by Brooks, Smith, Stone, and Tutte.

**Suggestions for reflection and for investigations**
**A 14.7:** Draw the network belonging to the rectangle in Fig. 14.5a (Fig. 14.5b).

## 14.6   A Game with Rectangular Dissections

The rectangles and squares, which can be dissected into squares of different sizes, can also be used for entertainment:

> Imagine that Fig. 14.10a represents the layout of a park, the lines are the paths (also the borderlines are paths) and the lengths in meters can be taken from Fig. 14.10b.
> You are standing at the bottom left corner and would like to take a walk as long as possible through the park or along the park, but without walking a segment twice. Who finds the longest way around the park?

The map shows the rectangular dissection of the ICM stamp. One preliminary consideration is that you can cover a particularly long distance by walking along large squares – as many as possible. Since the three largest squares are laying on the edge, the walk is mainly along the edge of the park.

The path entered in Fig. 14.10c has the following length:

$$77 + 2 \cdot 99 + 2 \cdot 78 + 57 + 3 \cdot 41 + 16 + 57 + 2 \cdot 21$$
$$+22 + 2 \cdot 43 + 2 \cdot 9 + 25 + 34 + 77 = 988$$

Altogether, the length of this path is almost 1 km.

> **Suggestions for reflection and for investigations**
> **A 14.8:** Determine the longest possible walking path through the park, the route map of which is shown in Fig. 14.1a, b (Fig. 14.2a, b).

**a**
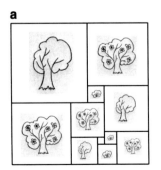

**b**
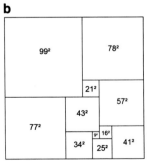

**c**
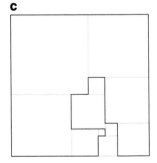

**Fig. 14.10   a–c** Map of a park with associated rectangular dissection and a proposal for a closed path around the park

## 14.7   References to Further Literature

The most comprehensive source of information on this topic is the website:

- https://www.squaring.net/index.html

In addition to detailed historical information, it contains extensive files with the graphics of *thousands* of rectangular and square tessellations.

On **Wikipedia** you will find further information and literature on the keyword in English (German, French):

- Squaring the square (Quadratur des Quadrats, Quadrature du carré)

Extensive informations can be found at **Wolfram Mathworld** under the keyword:

- Perfect square dissection

Further notes:
   More details on the history of squaring the square on the website of
   **The Trinity Mathematical Society**

- https://tms.soc.srcf.net/about-the-tms/the-squared-square/

The **Websites of Karl Scherer**

- https://karlscherer.com/prosqtre.html
- https://karlscherer.com/prosqtsq.html

deal with the tessellation of rectangles or squares by squares that do not necessarily differ in size; however, squares of the same size that are used several times must not have a common side *(nowhere-neat)* or no point in common *(no-touch)*.

# Kissing Circles

# 15

*Mathematics as a subject is so serious that no opportunity should be missed to make this subject more entertaining.*

*(Blaise Pascal, French mathematician and philosopher, 1623–1662)*

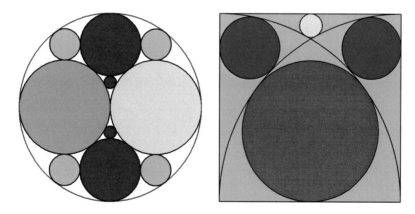

Among the challenging problems in geometry you often find the task of fitting in circles into other figures.

Many examples of this and of how to fit circles into other geometric figures can be found in Japanese temple geometry (Sangaku), which we will discuss in Sect. 15.6.

To solve these problems, one often needs knowledge of trigonometric theorems, especially the law of cosines and the law of sines.

© Springer-Verlag GmbH Germany, part of Springer Nature 2021
H. K. Strick, *Mathematics is Beautiful*, https://doi.org/10.1007/978-3-662-62689-4_15

## 15.1  Examination of Touching Circles using Trigonometric Methods

**Problem 1**

Given is a circle with center *M*, on whose diameter the centers *A*, *B* of two smaller circles are located. These two circles touch each other and touch the circle around *M*. We are looking for the two circles that touch the three given circles.

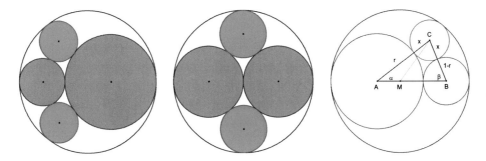

The figure on the left serves to illustrate the problem; the figure in the center shows the special case of two equally sized circles.

For reasons of symmetry, we consider only the circle around *C* above the diameter (see figure on the right).

By applying the law of cosines in the triangles *AMC* and *ABC*, you get:

$$\overline{MC}^2 = \overline{AM}^2 + \overline{AC}^2 - 2 \cdot \overline{AM} \cdot \overline{AC} \cdot \cos(\alpha)$$
$$\overline{BC}^2 = \overline{AB}^2 + \overline{AC}^2 - 2 \cdot \overline{AB} \cdot \overline{AC} \cdot \cos(\alpha)$$

The two equations can be solved for $2 \cdot \overline{AC} \cdot \cos\alpha$ so we get:

$$\left(\overline{AM}^2 + \overline{AC}^2 - \overline{MC}^2\right) \cdot \overline{AB} = \left(\overline{AB}^2 + \overline{AC}^2 - \overline{BC}^2\right) \cdot \overline{AM}$$

If one chooses 1 as radius of the outer circle and designates the radius of the left red circle with *r* and the radius of the circle to be determined (green) with *x*, that is

$$|AB| = 1, |AM| = 1 - r, |AC| = r + x, |MC| = 1 - x \text{ and } |BC| = 1 - r + x$$

we get the equation

$$1 \cdot \left[(1-r)^2 + (r+x)^2 - (1-x)^2\right] = (1-r) \cdot \left[1^2 + (r+x)^2 - (1-r+x)^2\right]$$

and from this

$$x = \frac{r \cdot (1-r)}{1 - r + r^2}.$$

With the help of this formula, it is, therefore, possible to calculate a radius $x$ for any radius $r$ ($0 < r < 1$), so that the circles with the radii $r$ and $1-r$ are lying symmetrical to a diameter of the circle with radius 1 and the circles with radii $r$, $1-r$, and $x$ are touching each other. Because of symmetry, another circle with radius $x$ can be added in the lower part of the figure.

**Examples**
In case $r = \frac{1}{2}$ (two circles of equal size lying symmetrically on the diameter), the radius of the complementary circle is $x = \frac{1}{3}$.
    In case $r = \frac{1}{3}$ (the right circle has a radius twice as large as the left circle) the radius of the complementary circle is $x = \frac{2}{7}$.

**Problem 2**
Given is a circle with center $M$, on whose diameter the center $A$ of a smaller circle is located, which touches the circle around $M$. We are looking for two circles that lie symmetrically to the diameter and touch the two given circles.

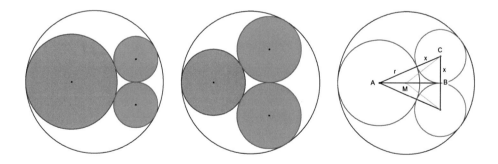

By applying the law of cosines in the triangle $AMC$, the following is obtained again:

$$\overline{MC}^2 = \overline{AM}^2 + \overline{AC}^2 - 2 \cdot \overline{AM} \cdot \overline{AC} \cdot \cos(\alpha)$$

If one chooses 1 as the radius of the outer circle and designates the radius of the red circle with $r$ and the radius of the circles to be determined (green) with $x$, then the approach results in

$$|AB| = 1, |AM| = 1 - r, |AC| = r + x, |MC| = 1 - x \text{ and } |BC| = x \text{ and}$$

$$\cos(\alpha) = \frac{(r + x)^2 + (1 - r)^2 - (1 - x)^2}{2 \cdot (r + x) \cdot (1 - r)}$$

The application of the law of sines in the right-angled triangle $ABC$ results in:

$$\sin(\alpha) = \frac{|BC|}{|AC|} = \frac{x}{r + x}$$

And using $\sin^2(a) + \cos^2(a) = 1$ we get after several transformations:

$$x = \frac{4r \cdot (1 - r)}{(1 + r)^2}$$

With the help of this formula, a radius $x$ can be determined for any radius $r$ ($0 < r < 1$) so that the circles with radius $r$ and radius $x$, which are lying symmetrically to a diameter, touch each other.

**Examples**
In case $r = \frac{2}{3}$ the radius of the two complementary circles is $x = 0.32$.

In the case, $r = 2 \cdot \sqrt{3} - 3 \approx 0.464$ the radius of the two complementary circles is as large as $x = 2 \cdot \sqrt{3} - 3 \approx 0.464$. The centers of the three inner circles define an equilateral triangle.

**Suggestions for reflection and for investigations**
**A 15.1:** Choose more examples to apply the solutions of problem 1 and 2 and draw these figures.

## 15.2 Descartes' Theorem

The formulas derived in Sect. 15.1 for determining the radii of touching circles are only special cases of a general theorem found by the French mathematician and philosopher René Descartes (1596–1650). The theorem fell into oblivion and was only rediscovered in the twentieth century by the Nobel Prize winner Frederick Soddy (1877–1956). In antiquity, the Greek mathematician Apollonius of Perga (262–190 B.C.) had discovered how such problems could be solved by means of a construction.

The following theorem was named after Descartes because he had in principle correctly recognized the connections between the radii; however, the equation he mentioned in a letter had a somewhat more complicated form.

As with the two problems above, the challenge is about three circles touching each other and we look for the radius of a circle that touches the three given circles.

**Theorem**
**Descartes' four-circle theorem**
If $r_1$, $r_2$, and $r_3$ are the radii of three circles touching each other and if $k_1$, $k_2$, $k_3$ are their reciprocal values (= called their curvatures), then the following is valid for the radius $r_4$ of the circle touching the three circles with $k_4 = \pm\frac{1}{r_4}$ is:

$$2 \cdot \left(k_1^2 + k_2^2 + k_3^2 + k_4^2\right) = \left(k_1 + k_2 + k_3 + k_4\right)^2 \tag{15.1}$$

◄

This equation can be used as follows: If the radii $r_1$, $r_2$, and $r_3$ of three circles touching each other are given, then these radii can be inserted into the equation and finally the equation is solved for the variable $r_4$ (or $k_4$).

Since this is a quadratic equation, the equation generally has two solutions, that is, there are two circles touching the three given circles. As can be seen in the following figure on the right, there is one circle touching the three circles from the outside, and one touching the three circles from the inside.

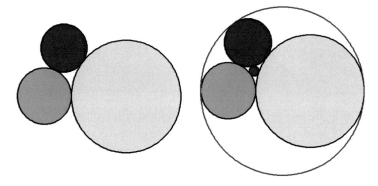

In the following, we refer to the curvature of the touching circles with $x$. It is worth solving the quadratic equation in general:

$$2 \cdot \left( k_1^2 + k_2^2 + k_3^2 + x^2 \right) = (k_1 + k_2 + k_3 + x)^2$$
$$2 \cdot \left( k_1^2 + k_2^2 + k_3^2 \right) + 2x^2 = (k_1 + k_2 + k_3)^2 + 2 \cdot (k_1 + k_2 + k_3) \cdot x + x^2$$

$$x^2 - 2 \cdot (k_1 + k_2 + k_3) \cdot x + 2 \cdot \left( k_1^2 + k_2^2 + k_3^2 \right) - (k_1 + k_2 + k_3)^2 = 0 \quad (15.2)$$

$$[x - (k_1 + k_2 + k_3)]^2 = 2 \cdot \left[ (k_1 + k_2 + k_3)^2 - \left( k_1^2 + k_2^2 + k_3^2 \right) \right] \quad (15.3)$$

**Example 1**

Given are three circles with the radii $r_1 = 15, r_2 = 15, r_3 = 10$.

Starting from formula (15.3) it results:

$$\left[ x - \left( \frac{1}{15} + \frac{1}{15} + \frac{1}{10} \right) \right]^2 = 2 \cdot \left[ \left( \frac{1}{15} + \frac{1}{15} + \frac{1}{10} \right)^2 - \left( \frac{1}{225} + \frac{1}{225} + \frac{1}{100} \right) \right]$$

$$\Leftrightarrow \left[ x - \frac{7}{30} \right]^2 = 2 \cdot \left[ \left( \frac{7}{30} \right)^2 - \frac{17}{900} \right] \Leftrightarrow \left( x - \frac{7}{30} \right)^2 = \frac{64}{900} \Leftrightarrow x = \frac{1}{2} \lor x = -\frac{1}{30}$$

The quadratic equation has two solutions, which can be interpreted as follows:

The radius of the red colored circle touching from the inside, is $r_4 = 2$ and that of the circle touching from the outside (hence the negative sign) $r_5 = 30$.

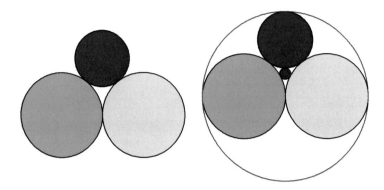

Since in this example the radii $r_1$ and $r_2$ are equal and exactly half the size of the radius $r_5$ of the outer circle, the center of the outer circle is located at the point of contact of the green and yellow colored circles. Therefore, in this case, the figure can be completed symmetrically.

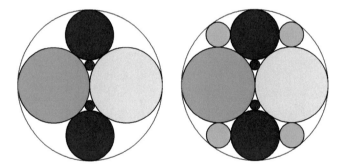

In this figure, we can now fit in more circles, for example, to the top left. We are looking for a circle with radius $r_6$, which touches the upper blue colored circle with radius $r_3 = 10$, the green colored circle with radius $r_1 = 15$, and the outer red circle with radius $r_5 = 30$.

But the yellow colored circle also has this property, that is we already know one of the expected two solutions of the quadratic equation. This makes the solution of the quadratic equation much easier.

The other solution is obtained by a simple calculation using Vieta's formula (Vieta is the latin form of the name of François Viète, 1540–1603):

---

**Theorem**

**Vieta's formula**

*If a quadratic equation $x^2 + px + q = 0$ has the solutions $x_1$ and $x_2$ then we have:*

$$x_1 + x_2 = -p \text{ and } x_1 \cdot x_2 = q$$

◀

When we compare these properties with the formula (15.2), we have:

$$x_1 + x_2 = 2 \cdot (k_1 + k_2 + k_3) \tag{15.4}$$

**Example 1 (continued)**

In this example, the curvatures $k_{green} = \frac{1}{15}, k_{blue} = \frac{1}{10}$ and $k_{outside} = -\frac{1}{30}$ are known and $x_2 = k_{yellow} = \frac{1}{15}$.

Therefore, the curvature of the light blue colored circles results for the wanted curvature:

$$x_1 = 2 \cdot \left( \frac{1}{15} + \frac{1}{10} - \frac{1}{30} \right) - \frac{1}{15} = \frac{6}{30} = \frac{1}{5}$$

That means that these circles have the radius $r_6 = 5$.

The procedure just applied can now be continued arbitrarily to determine further radii of circles by which the existing gaps can be filled.

The radii of the circles, which will result of the next steps, can be seen in Fig. 15.1.

This results in a sequence of **kissing circles.** In Fig. 15.2 the outer circle has the radius 1 (instead of 30); the integral values labeled in the circles therefore correspond to the curvatures.

*Note:* The example also includes problem 1 from Sect. 15.1.

| Wanted: two circles, which touch the three following circles | Radius already known | $x_1 = 2 \cdot (k_1 + k_2 + k_3) - x_2$ | Newly determined radius |
|---|---|---|---|
| light blue (5), blue (10), outer (30) | green (15) | $2 \cdot \left( \frac{1}{5} + \frac{1}{10} - \frac{1}{30} \right) - \frac{1}{15} = \frac{14}{30}$ | $\frac{30}{14}$ |
| light blue (5), green (15), outside (30) | blue (10) | $2 \cdot \left( \frac{1}{5} + \frac{1}{15} - \frac{1}{30} \right) - \frac{1}{10} = \frac{11}{30}$ | $\frac{30}{11}$ |
| light blue (5), green (15), blue (10) | outside (30) | $2 \cdot \left( \frac{1}{5} + \frac{1}{15} + \frac{1}{10} \right) + \frac{1}{30} = \frac{23}{30}$ | $\frac{30}{23}$ |
| green (15), blue (10), red (2) | yellow (15) | $2 \cdot \left( \frac{1}{15} + \frac{1}{10} + \frac{1}{2} \right) - \frac{1}{15} = \frac{38}{30}$ | $\frac{30}{38}$ |
| green (15), yellow (15), red (2) | blue (10) | $2 \cdot \left( \frac{1}{15} + \frac{1}{15} + \frac{1}{2} \right) - \frac{1}{10} = \frac{35}{30}$ | $\frac{30}{35}$ |

**Fig. 15.1** Radii, which in example 1 result from repeated application of Vieta's formula

**Fig. 15.2** Sequence of
the kissing circles for
$r_1 = r_2 = \frac{1}{2}$ and $r_3 = \frac{1}{3}$ (with
kind permission of the author
David Austin and the American
Mathematical Society)

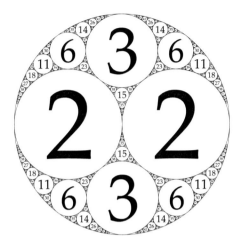

---

**Suggestions for reflection and for investigations**

**A 15.2:** Show that the two formulas resulting from the two problems in Sect. 15.1
are just special cases of Descartes' theorem.

---

## 15.3  Examples with Integral Radii

The fact that only integral curvatures occur in Fig. 15.4 is due to the equation from
formula (15.4):

$$x_1 + x_2 = 2 \cdot (k_1 + k_2 + k_3)$$

If you start with integral curvatures $k_1, k_2, k_3$ and if in the first step, that is, when solving
the quadratic equation, integral curvatures $k_4$ and $k_5$ are obtained again then all further
curvatures must be integral in the following steps.

But, when are the curvatures $k_4$ and $k_5$ integers?

A further transformation of the quadratic equation in formula (15.3) results in:
$[x - (k_1 + k_2 + k_3)]^2 = 2 \cdot (2k_1k_2 + 2k_1k_3 + 2k_2k_3)$, that is:

$$x = (k_1 + k_2 + k_3) \pm 2 \cdot \sqrt{k_1k_2 + k_1k_3 + k_2k_3} \qquad (15.5)$$

So whether the curvatures $k_4$ and $k_5$ are integers, too, depends on the square root.

---

**Rule**

**Determination of examples with integral radii**

If the sum of products $k_1k_2 + k_1k_3 + k_2k_3$ of the circular curvatures $k_1, k_2, k_3$ is a
square number, you get touching circles with all integral curvatures. If one then
chooses the smallest common multiple of the curvatures as the radius of the surrounding circle (outside), then the radii of the inner circles are also integral. ◀

**Example 1 (continued)**

Here this condition is fulfilled with $k_1 = k_2 = 2$ and $k_3 = 3$:

$k_1 k_2 + k_1 k_3 + k_2 k_3 = 2 \cdot 2 + 2 \cdot 3 + 2 \cdot 3 = 16$ is a square number,

and it results in: $k_4 = 15$ and $k_5 = -1$.

Since 30 is the smallest common multiple of the curvatures 2, 3, and 15, the radii of the inner circles are integers, namely $r_1 = \frac{30}{k_1} = \frac{30}{2} = 15 = r_2$, $r_3 = \frac{30}{k_3} = \frac{30}{3} = 10$, and $r_4 = \frac{30}{k_4} = \frac{30}{15} = 2$.

**Example 2**

There are three circles with the radii $r_1 = 7$ (green), $r_2 = 14$ (yellow), $r_3 = 6$ (blue).

The radii of the two touching circles are determined by means of formula (15.5):

$$x = \tfrac{1}{7} + \tfrac{1}{14} + \tfrac{1}{6} \pm 2 \cdot \sqrt{\frac{1}{7 \cdot 14} + \frac{1}{7 \cdot 6} + \frac{1}{14 \cdot 6}} = \frac{6 + 3 + 7}{42} \pm 2 \cdot \sqrt{\frac{18 + 42 + 21}{7^2 \cdot 2^2 \cdot 3^2}} = \frac{16}{42} \pm 2 \cdot \frac{9}{42}$$

That is $x = \frac{17}{21} \vee x = -\frac{1}{21}$.

The two touching circles, therefore, have the radii $r_4 = \frac{21}{17}$ and $r_5 = 21$.

Also for this example, one notices that the radius of the outer circle is equal to the sum of the radii of two of the initial circles. Therefore, this figure can also be completed symmetrically without further calculations.

As in example 1, you can then use formula (15.4) to determine the radii of the closest adjacent circles (see the following table and Fig. 15.3).

| Wanted: two circles, which touch the three following circles | Radius already known | $x_1 = 2 \cdot (k_1 + k_2 + k_3) - x_2$ | Newly determined radius |
|---|---|---|---|
| green (7), blue (6), outside (21) | yellow (14) | $2 \cdot \left( \frac{1}{7} + \frac{1}{6} - \frac{1}{21} \right) - \frac{1}{14} = \frac{19}{42}$ | $\frac{42}{19}$ (light blue) |
| blue (6), yellow (14), outside (30) | green (7) | $2 \cdot \left( \frac{1}{6} + \frac{1}{14} - \frac{1}{21} \right) - \frac{1}{7} = \frac{10}{42}$ | $\frac{42}{10}$ (pink) |

In Fig. 15.3 the radius of the outer circle is equal to $\frac{1}{2}$. This results in the curvatures

$$k_1 = \frac{42}{7} = 6; \quad k_2 = \frac{42}{14} = 3; \quad k_3 = \frac{42}{6} = 7; \quad k_4 = \frac{42}{\frac{21}{17}} = 34 \text{ and so on.}$$

a                                          b

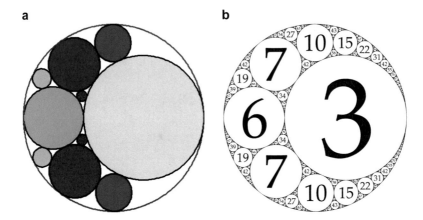

**Fig. 15.3  a, b** Sequence of the kissing circles in example 2 (with kind permission of the author David Austin and the American Mathematical Society)

By systematic trial and error, it is possible to find other examples with integral curvatures using formula (15.5) (see the table in Fig. 15.4; the first two rows contain examples 1 and 2).

**Suggestions for reflection and for investigations**

**A 15.3:** Calculate the curvatures of those circles that can be calculated in the next step starting with the examples in the table in Fig. 15.4. Specify a suitable integral radius for the outer circle. Draw the figures.

**A 15.4:** Determine the radii of the circles in the following illustrations:

1. $r_1 = r_2 = r_3 = 1$ (green, yellow, blue)
2. $r_1 = r_2 = 5$ (green, yellow), $r_3 = 8$ (blue)
3. $r_1 = 6$ (green), $r_2 = 5$ (yellow), $r_3 = 3$ (blue)

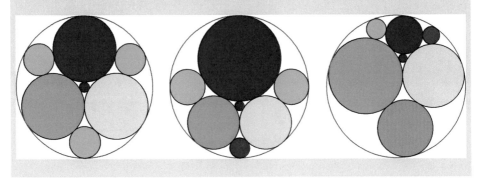

| $k_1$ | $k_2$ | $k_3$ | $k_1 + k_2 + k_3$ | $k_1k_2 + k_1k_3 + k_2k_3$ | $k_4$ | $k_5$ |
|---|---|---|---|---|---|---|
| 2 | 2 | 3 | 7 | 16 | 15 | -1 |
| 3 | 6 | 7 | 16 | 81 | 34 | -2 |
| 5 | 8 | 8 | 21 | 144 | 45 | -3 |
| 1 | 1 | 4 | 6 | 9 | 12 | 0 |
| 1 | 4 | 9 | 14 | 49 | 28 | 0 |
| 2 | 3 | 6 | 11 | 36 | 23 | -1 |
| 3 | 7 | 10 | 20 | 121 | 42 | -2 |
| 8 | 9 | 9 | 26 | 225 | 56 | -4 |

**Fig. 15.4**  Examples of integral curvatures

**A 15.5:** Analyze the following illustrations and reconstruct the first steps of calculating the radii of the figures. (Reprinted with kind permission of the author David Austin and the American Mathematical Society).

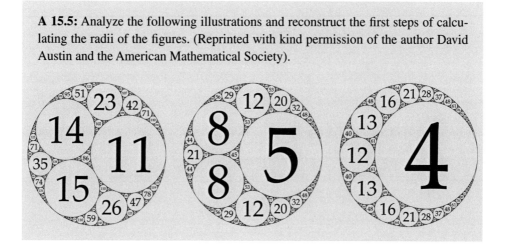

## 15.4  Pappus chains

In the above graphics in Fig. 15.2 and 15.3 by David Austin, regularities are noticeable with regard to the curvatures of the circles that form a "chain" around the initial circles:

- Example 1: 2, 3, 6, 11, 18, 27, 38, …
- Example 2: 6, 7, 10, 15, 22, 31, 42, …

Each is an infinite sequence of circles with decreasing radii, which are called **Pappus chains** after their discoverer, the Greek mathematician **Pappus of Alexandria,** who lived around the year 320 A.D. The curvatures form an arithmetic sequence of the 2nd order (the second sequence of differences is constant), see more on this in Chapt. 16.

In example 1 it looks like this:

| 2 | | 3 | | 6 | | 11 | | 18 | | 27 | | 38 | | ... |
|---|---|---|---|---|---|----|---|----|---|----|---|----|---|-----|
| | 1 | | 3 | | 5 | | 7 | | 9 | | 11 | | ... | |
| | | 2 | | 2 | | 2 | | 2 | | 2 | | ... | | |

In the following, the first elements of a Pappus chain are analyzed using Descartes' theorem.

The radius of the outer circle $K$ in Fig. 15.5 is denoted by $r$, the radius of the circle $K_0$, which is left uncolored, by $r_0$. Then the radius of the circle $K_0'$ is equal to $r - r_0$.

For the radius $r_1$ of the two circles that touch symmetrically above and below, we have – according to Descartes' theorem:

$$\frac{1}{r_1} + \frac{1}{r_1} = 2 \cdot \left( \frac{1}{r_0} + \frac{1}{r - r_0} - \frac{1}{r} \right),$$

so

$$\frac{1}{r_1} = \frac{1}{r_0} + \frac{1}{r - r_0} - \frac{1}{r}.$$

For the "next" circle $K_2$ of the chain the following is valid: $K_2$ touches $K_1$ and $K_0$ and is touched by $K$ from the outside.

But $K$, $K_0$, and $K_1$ also touch the circle $K_0'$. Therefore from the circle theorem we get:

$$\frac{1}{r_2} + \frac{1}{1 - r_0} = 2 \cdot \left( \frac{1}{r_0} + \frac{1}{r_1} - \frac{1}{r} \right)$$

**Fig. 15.5** Schema of a Pappus chain (labeled graphic taken from the English Wikipedia article *Pappus chain*)

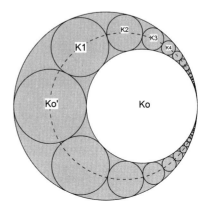

Replacing the term for $\frac{1}{r_1}$ and solving it for $\frac{1}{r_2}$, therefore, this results in:

$$\frac{1}{r_2} = 2 \cdot \left( \frac{1}{r_0} + \left( \frac{1}{r_0} + \frac{1}{r - r_0} - \frac{1}{r} \right) - \frac{1}{r} \right) - \frac{1}{r - r_0} = \frac{4}{r_0} + \frac{1}{r - r_0} - \frac{4}{r}$$

The following applies to the "next" circle $K_3$ of the chain: $K_3$ touches $K_2$ and $K_0$, and is also touched by $K$ from the outside. But $K$, $K_0$ and $K_2$ also touch circle $K_1$. From Descartes' theorem, we get:

$$\frac{1}{r_3} + \frac{1}{r_1} = 2 \cdot \left( \frac{1}{r_0} + \frac{1}{r_2} - \frac{1}{r} \right)$$

Replace the terms for $\frac{1}{r_1}$ and $\frac{1}{r_2}$ and resolving it for $\frac{1}{r_3}$, this results in:

$$\frac{1}{r_3} = 2 \cdot \left( \frac{1}{r_0} + \left( \frac{4}{r_0} + \frac{1}{r - r_0} - \frac{4}{r} \right) - \frac{1}{r} \right) - \left( \frac{1}{r_0} + \frac{1}{r - r_0} - \frac{1}{r} \right) \text{ well } \frac{1}{r_3} = \frac{9}{r_0} + \frac{1}{r - r_0} - \frac{9}{r}.$$

Analogously we get in the next step:

$$\frac{1}{r_4} + \frac{1}{r_2} = 2 \cdot \left( \frac{1}{r_0} + \frac{1}{r_3} - \frac{1}{r} \right) \text{ and further } \frac{1}{r_4} = \frac{16}{r_0} + \frac{1}{r - r_0} - \frac{16}{r}.$$

The general principle has now become clear.

For the circle $K_{n+1}$ of the chain, following as far as the circle $K_n$, we get:

Circle $K_{n+1}$ touches circle $K_n$ and circle $K_0$, in addition, it is touched from the outside by circle $K$. Circle $K_{n-1}$ is also touched by $K$, $K_0$, and $K_n$. Therefore, the following relationship results from the circle theorem:

$$\frac{1}{r_{n+1}} + \frac{1}{r_{n-1}} = 2 \cdot \left( \frac{1}{r_0} + \frac{1}{r_n} - \frac{1}{r} \right)$$

After transforming this finally we get:

---

**Theorem**

**Radii of the circles in a Pappus chain**

For the radius of the $(n+1)$-st circle of a Pappus chain we have:

$$\frac{1}{r_{n+1}} = \frac{(n+1)^2}{r_0} + \frac{1}{r - r_0} - \frac{(n+1)^2}{r} \qquad (15.6)$$

◀

---

**Example 1: Radii of a Pappus chain**

The calculation of the radii is as follows:

Here $r = 1$ and $r_0 = \frac{1}{2}$.

For the curvature of the two adjacent circles $K_1$ results:

$$\frac{1}{r_1} = \frac{1}{\frac{1}{2}} + \frac{1}{\frac{1}{2}} - \frac{1}{1} = 3$$

Further it follows for the next circles

$$\frac{1}{r_2} = \frac{4}{\frac{1}{2}} + \frac{1}{\frac{1}{2}} - \frac{4}{1} = 6, \frac{1}{r_3} = \frac{9}{\frac{1}{2}} + \frac{1}{\frac{1}{2}} - \frac{9}{1} = 11, \frac{1}{r_4} = \frac{16}{\frac{1}{2}} + \frac{1}{\frac{1}{2}} - \frac{16}{1} = 18$$

and in general:

$$\frac{1}{r_{n+1}} = \frac{(n+1)^2}{\frac{1}{2}} + \frac{1}{\frac{1}{2}} - \frac{(n+1)^2}{1} = (n+1)^2 + 2 = n^2 + 2n + 3$$

When we insert $n = 1, 2, 3, \ldots$ we get the sequence of the curvatures 2, 3, 6, 11, 18, 27, 38, … (see Fig. 15.2).

**Suggestions for reflection and for investigations**

**A 15.6:** Perform the calculation of the radii for **example 2**.

   Show: $\frac{1}{r_{n+1}} = n^2 + 2n + 7$.

**A 15.7:** Determine the term for calculating the radii of the chain for one of the Pappus chains in the figures of **A 15.5**.

## 15.5   Touching circles with curvature 0

In the table of Fig. 15.4, the triples with solution $k_5 = 0$ are noticeable. Circles with curvature 0 have an infinitely large radius – these are straight lines that touch the circles, see the following figures.

The starting triple (1, 1, 4) includes $r_1 = 12 \cdot 1 = 12$ (green), $r_2 = 12 \cdot 1 = 12$ (yellow), and $r_3 = 12 \cdot \frac{1}{4} = 3$ (blue). With formula (15.5) it results: $r_4 = 12 \cdot \frac{1}{12} = 1$ (red) and $r_5 = 12 \cdot \frac{1}{0} = \infty$.

The starting triple (1, 4, 9) includes $r_1 = 28 \cdot 1 = 28$ (green), $r_2 = 28 \cdot \frac{1}{4} = 7$ (yellow), and $r_3 = 28 \cdot \frac{1}{9} = \frac{28}{9}$ (blue). With formula (15.5) the following results: $r_4 = 28 \cdot \frac{1}{28} = 1$ (red) and $r_5 = 28 \cdot \frac{1}{0} = \infty$.

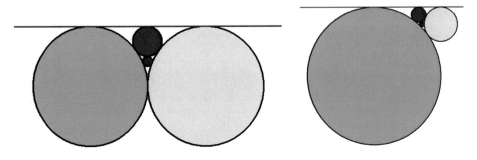

| $k_1$ | $k_2$ | $k_3$ | $k_1 + k_2 + k_3$ | $k_1k_2 + k_1k_3 + k_2k_3$ | $k_4$ | $k_5$ |
|---|---|---|---|---|---|---|
| 1 | 1 | 4 | 6 | 9 | 12 | 0 |
| 1 | 4 | 9 | 14 | 49 | 28 | 0 |
| 1 | 9 | 16 | 26 | 169 | 52 | 0 |
| 1 | 16 | 25 | 42 | 441 | 84 | 0 |
| 1 | 25 | 36 | 62 | 961 | 124 | 0 |
| 1 | 36 | 49 | 86 | 1849 | 172 | 0 |
| 1 | 49 | 64 | 114 | 3249 | 228 | 0 |
| 1 | 64 | 81 | 146 | 5329 | 292 | 0 |
| 4 | 9 | 25 | 38 | 361 | 76 | 0 |
| 4 | 25 | 49 | 78 | 1521 | 156 | 0 |
| 4 | 49 | 81 | 134 | 4489 | 268 | 0 |
| 9 | 16 | 49 | 74 | 1369 | 148 | 0 |
| 9 | 25 | 64 | 98 | 2401 | 196 | 0 |

**Fig. 15.6**  Examples of touching circles with curvature 0

General consideration of the examples with curvature 0 lead to the following:

From formula (15.1) it follows that $2 \cdot \left(k_1^2 + k_2^2 + k_3^2 + 0^2\right) = (k_1 + k_2 + k_3 + 0)^2$ and further from formula (15.5):

$$0 = (k_1 + k_2 + k_3) - 2 \cdot \sqrt{k_1k_2 + k_1k_3 + k_2k_3}$$

Both can be combined to give:

$$k_1^2 + k_2^2 + k_3^2 = 2 \cdot (k_1k_2 + k_1k_3 + k_2k_3)$$

In the table of Fig. 15.6 further examples are given, which can be found by systematic trial and error; multiples of triples have been omitted.

It is striking that all examples marked with $k_5 = 0$ consist of only square numbers.

On the basis of the examples recorded, it is reasonable to assume the following rule:

---

**Rule**

**Common tangent for three touching circles**

For all natural numbers $n$, $k$ we get:

If the curvatures of three circles touching each other can be displayed in the form $(n^2; (n+k)^2; (2n+k)^2)$, then there is a straight line that touches the three circles, which forms a common tangent for the three circles. ◀

The proof succeeds by calculating the terms on both sides of the equation; in fact, the following applies

$$n^4 + (n+k)^4 + (2n+k)^4 = 2 \cdot \left[ n^2 \cdot (n+k)^2 + n^2 \cdot (2n+k)^2 + (n+k)^2 + (2n+k)^2 \right]$$
$$= 18 \cdot n^4 + 36 \cdot n^3 \cdot k + 30 \cdot n^2 \cdot k^2 + 12 \cdot n \cdot k^3 + 2 \cdot k^4$$

**Suggestions for reflection and for investigations**
**A 15.8:** What is the curvature of a circle that touches three touching circles with a common tangent?
**A 15.9:** What are the curvatures of the circles that touch two touching circles and the common tangent?

## 15.6   Sangaku

In the following, not only circles but also other geometric figures are to be considered (especially semicircles, squares, and equilateral triangles), and circles are to be fitted into these figures. As mentioned above, many examples of this kind can be found in Japanese temple geometry (Sangaku).

The Japanese word *Sangaku* means something like "mathematical tablets". These colored wooden tablets depicting a geometrical problem were hung in temples by pilgrims in Japan in earlier centuries – both as offerings to the deity and as "puzzles" for other visitors.

The famous Japanese mathematician Seki Kowa (1642–1708), a contemporary of Newton and Leibniz, also worked on Sangaku problems.

**Example 1**
Into a semicircle, two circles of the same size are drawn (green) so that they touch each other and also the arc and the diameter of the semicircle. Further, circles are fitted into the spaces above (yellow) and below (blue), and left and right (red).

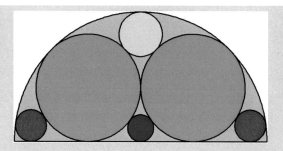

**Solution:**

For the radius of the semicircle we choose the length 1.

Determination of the radius $r_1$ of the green-colored circles:

A square with side length $r_1$ is defined by the center of the semicircle and the center of a green-colored circle, the length of its diagonal is, according to the Pythagorean theorem, $r_1 \cdot \sqrt{2}$.

Thus the following applies to the radius of the semicircle

$$r_1 \cdot \sqrt{2} + r_1 = r_1 \cdot \left( \sqrt{2} + 1 \right) = 1 \text{ that is}$$

$$r_1 = \frac{1}{\sqrt{2}+1} = \frac{\sqrt{2}-1}{\left(\sqrt{2}+1\right)\left(\sqrt{2}-1\right)} = \frac{\sqrt{2}-1}{2-1} = \sqrt{2} - 1 \approx 0.4142.$$

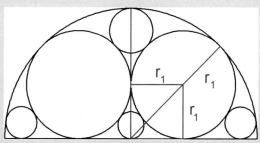

Determination of the radius $r_2$ of the yellow-colored circle:

According to the theorem of Pythagoras, in the right-angled triangle with the shorter sides of length $r_1$ and $1 - r_1 - r_2$, and the hypotenuse of length $r_1 + r_2$:

$r_1^2 + (1 - r_1 - r_2)^2 = (r_1 + r_2)^2$ that is

$r_1^2 + 1 + r_1^2 + r_2^2 - 2r_1 - 2r_2 + 2r_1r_2 = r_1^2 + 2r_1r_2 + r_2^2$ and, therefore,

$2r_2 = 1 + r_1^2 - 2r_1 = (1 - r_1)^2.$

Insertion of $r_1 = \sqrt{2} - 1$ results in $r_2 = 3 - 2\sqrt{2} \approx 0.1716.$

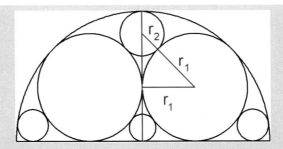

Determination of the radius $r_3$ of the red-colored circles:

To determine this radius, you need the horizontal distance $x$ between the center of one of the green circles and the center of the red circle concerned. In the two right-angled triangles of the auxiliary figure we have:

red triangle: $x^2 + (r_1 - r_3)^2 = (r_1 + r_3)^2$ that is $x^2 = 4r_1r_3$,

blue triangle: $(r_1 + x)^2 + r_3^2 = (1 - r_3)^2$.

If one replaces in the second equation $r_3$ by $\frac{x^2}{4r_1}$ then one obtains after transformation:

$(2r_1 + 1) \cdot x^2 + 4r_1^2 x = 2r_1 \cdot (1 - r_1^2)$.

Insertion of $r_1 = \sqrt{2} - 1$ that is $r_1^2 = 3 - 2\sqrt{2}$, and $1 - r_1^2 = 2\sqrt{2} - 2$ leads to a quadratic equation whose positive solution $x \approx 0.4531$ is.

It follows: $r_3 \approx 0.1239$.

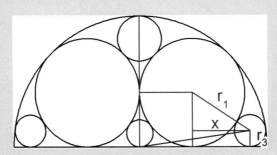

Determination of the radius $r_4$ of the blue-colored circle:

For the right-angled triangle in the following figure we get: $r_1^2 + (r_1 - r_4)^2 = (r_1 + r_4)^2$.

After forming, this results in $r_1 = 4r_4$ that is

$r_4 = \frac{1}{4} \cdot r_1 = \frac{1}{2} - \frac{1}{4} \cdot \sqrt{2} \approx 0.1036$.

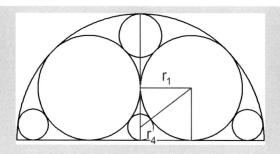

**Suggestions for reflection and for investigations**

**A 15.10:** Show that for the radii of the circles in the following semicircle with radius 1 we get:

$$r_{yellow} = \tfrac{1}{2}, r_{green} = \tfrac{1}{4}, r_{blue} \approx 0.128, r_{red} \approx 0.073.$$

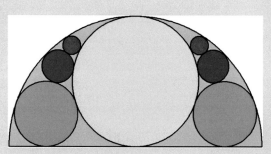

**Example 2**

The following figure contains five circles (gold, blue, red), which are fitted into a square with two semicircles – but we only have to do a calculation for the red-colored circles.

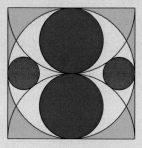

**Solution:**

To simplify the calculation we choose 2 as the length of the square side. In the following auxiliary figure, a right-angled triangle can then be seen, in which the theorem of Pythagoras can be applied. Thus we get:

$(1-r)^2 + 1^2 = (1+r)^2$ that is $1 - 2r + r^2 + 1 = 1 + 2r + r^2$ and therefore, $4r = 1$ that is $r = \frac{1}{4}$.

So the red-colored circles have a radius that is one-eighth of the side length of the square.

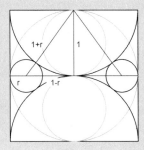

**Suggestions for reflection and for investigations**

**A 15.11:** Determine the radii of the circles enclosed in the outer figures.

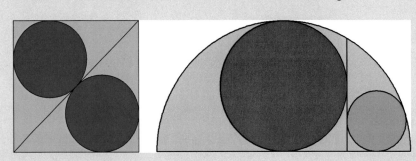

**A 15.12:** Show that for the radii of the circles in the following square figure (side length 1) we get:

$r_1 = r_{yellow} \approx 0.414$ and $r_2 = r_{red} \approx 0.172$.

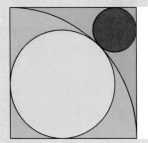
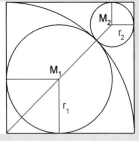

**A 15.13:** Determine the side length of the enclosed square (light yellow) and the radii of the two circles (after a Sangaku from 1820).

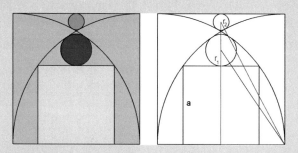

**A 15.14:** Show that for the radii of the circles in the following square (side length 1) we get:

$r_1 = r_{red} \approx 0.293$ and $r_2 = r_{green} \approx 0.160$. For the auxiliary quantity $x$, we get: $x \approx 0.180$.

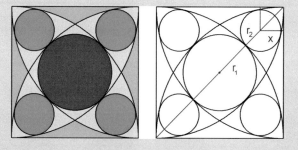

**A 15.15:** Show that for the radii of the circles in the following square (side length 1) we get:

$r_1 = r_{red} = \frac{3}{8} = 0.375$, $r_2 = r_{blue} = \frac{1}{6} \approx 0.167$, and $r_3 = r_{gold} = \frac{1}{16} = 0.0625$.

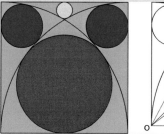
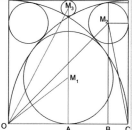

**A 15.16:** For the following square (side length 1), investigate how the total area $A$ of the green and gold-colored circles depends on the radii. For which radii has $A$ a maximum, for which a minimum? What is the radius for circles of equal size, see third figure (right)?

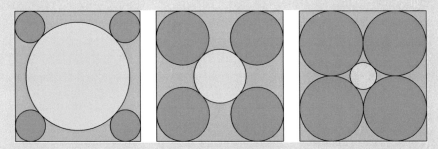

**A 15.17:** For the following equilateral triangle (side length 1), examine how the total area $A$ of the red or orange-colored circles depends on the radii. For which radii has $A$ a maximum, for which a minimum? What is the size of the radius in the case of circles of equal size, see third figure (right)?

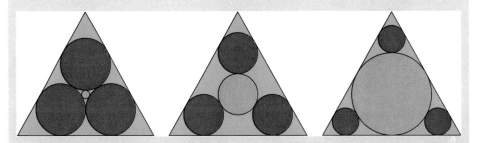

**A 15.18:** How must the outer square be divided so that the five enclosed circles are of the same size (according to a Sangaku from 1811)?

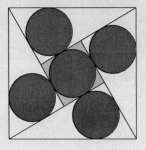

**A 15.19:** Determine the radii of the circles and arcs.

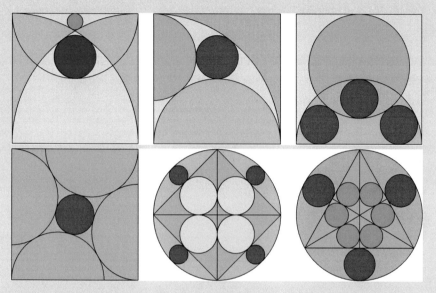

**A 15.20:** Analyze the following figure consisting of 4 circles.

Show: The segment AB is divided by the upper intersection point of the two inner circles in the ratio of the golden section.

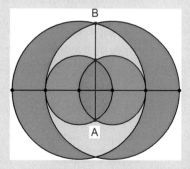

## 15.7    References to Further Literature

On **Wikipedia** you can find further information and literature on the keywords in English (German, French):

- Descartes' theorem (Satz von Descartes, Théorème de Descartes),
- Problem of Apollonius (Apollonisches Problem, Problème des contacts),
- Pappus chain (Pappos-Kette, –)
- Sangaku (Sangaku, Sangaku)

Extensive informations can be found at **Wolfram Mathworld** under the keywords:

- Descartes Circle Theorem, Apollonius Problem, Pappus Chain, Sangaku Problem

A detailed description, also of the connections with complex numbers, which could not be dealt with here, can be found in:

- Austin, David, *When Kissing Involves Trigonometry,* https://www.ams.org/samplings/feature-column/fcarc-kissing
- Lagarias, Jeffrey C. et al. (2002), *Beyond the Descartes circle theorem,* American Mathematical Monthly 109, S. 338–361.

On the subject of Sangaku the following book is recommended

- Huvent, Géry (2008): Sangaku – le mystère des énigmes géométriques japonaises, Dunod, Paris

and the websites

- www.zum.de/Faecher/Materialien/rubin/sangaku.html
- www.cut-the-knot.org/pythagoras/Sangaku.shtml
- www.matheraetsel.de/sangaku.html
- u.osu.edu/unger.26/online-publications/sangaku-problems-involving-ellipses/

# Sums of Powers of Consecutive Natural Numbers

# 16

*The good Lord made the integers, all else is the work of men.*

*(Leopold Kronecker, German mathematician, 1823–1891)*

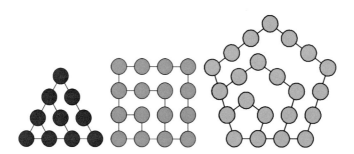

In Chap. 2 we explained how to calculate the sum of the first $n$ natural numbers in a descriptive way. Also, some ideas were presented, from which formulas for calculating the sum of the first $n$ square numbers or cube numbers can be derived.

In this chapter, further methods for calculating the sum of the powers of natural numbers will be presented, which have been discovered and developed over the centuries by various mathematicians. One of the first mathematicians of modern times to work with these formulas was Johannes Faulhaber (1580–1635). Following a suggestion by the American computer scientist Donald E. Knuth, the sum formulas are often referred to as **Faulhaber's formulas.**

© Springer-Verlag GmbH Germany, part of Springer Nature 2021

H. K. Strick, *Mathematics is Beautiful*, https://doi.org/10.1007/978-3-662-62689-4_16

**Sums of powers of the first $n$ natural numbers**

$$1 + 2 + 3 + \ldots + n = \tfrac{1}{2} \cdot n^2 + \tfrac{1}{2} \cdot n = \tfrac{1}{2} \cdot n \cdot (n+1) \tag{16.1}$$

$$1^2 + 2^2 + 3^2 + \ldots + n^2 = \tfrac{1}{3}n^3 + \tfrac{1}{2}n^2 + \tfrac{1}{6}n = \tfrac{1}{6} \cdot n \cdot (n+1) \cdot (2n+1) \tag{16.2}$$

$$1^3 + 2^3 + 3^3 + \ldots + n^3 = \tfrac{1}{4}n^4 + \tfrac{1}{2}n^3 + \tfrac{1}{4}n^2 = \left[\tfrac{1}{2} \cdot n \cdot (n+1)\right]^2 \tag{16.3}$$

$$1^4 + 2^4 + 3^4 + \ldots + n^4 = \tfrac{1}{5}n^5 + \tfrac{1}{2}n^4 + \tfrac{1}{3}n^3 - \tfrac{1}{30}n$$
$$= \tfrac{1}{30} \cdot n \cdot (n+1) \cdot (2n+1) \cdot \left(3n^2 + 3n - 1\right) \tag{16.4}$$

$$1^5 + 2^5 + 3^5 + \ldots + n^5 = \tfrac{1}{6}n^6 + \tfrac{1}{2}n^5 + \tfrac{5}{12}n^4 - \tfrac{1}{12}n^2$$
$$= \tfrac{1}{12} \cdot n^2 \cdot (n+1)^2 \cdot \left(2n^2 + 2n - 1\right) \tag{16.5}$$

◀

In his works *Miracula Arithmetica* (1622) and *Academia Algebrae* (1631) Johann Faulhaber developed sum formulas for the powers of the first $n$ natural numbers up to the twelfth power.

For this purpose he systematically created tables with the first elements of the sequence of natural numbers and their powers, then the corresponding series and also columns in which he defined multiples of these elements.

By "looking" directly at the tables, one can see that there is a connection $\Sigma k^3 = (\Sigma k)^2$ between the sum of the first $n$ cube numbers (reduced notation: $\Sigma k^3$ for $1^3 + 2^3 + 3^3 + \ldots + n^3$) and the sum of the first $n$ natural numbers (reduced notation: $\Sigma k$ for $1 + 2 + 3 + \ldots + n$) (see also formula 16.3).

| $n$ | $n^2$ | $n^3$ | $\Sigma k$ | $\Sigma k^2$ | $\Sigma k^3$ | $n \cdot \Sigma k$ | $n \cdot \Sigma k^2$ |
|-----|-------|-------|------------|--------------|--------------|---------------------|----------------------|
| 1 | 1 | 1 | 1 | 1 | 1 | 1 | 1 |
| 2 | 4 | 8 | 3 | 5 | 9 | 6 | 10 |
| 3 | 9 | 27 | 6 | 14 | 36 | 18 | 42 |
| 4 | 16 | 64 | 10 | 30 | 100 | 40 | 120 |
| 5 | 25 | 125 | 15 | 55 | 225 | 75 | 275 |
| 6 | 36 | 216 | 21 | 91 | 441 | 126 | 546 |
| 7 | 49 | 343 | 28 | 140 | 784 | 196 | 980 |
| 8 | 64 | 512 | 36 | 204 | 1296 | 288 | 1632 |

He also discovered that there is a "simple" relationship between the sum $\Sigma k^2 = 1^2 + 2^2 + 3^2 + \ldots + n^2$ of the first $n$ square numbers and the sum $\Sigma k$ of the first $n$ natural numbers: From the approach $\Sigma k^2 = (a \cdot n + b) \cdot (\Sigma k)$ one obtains the coefficients $a = \frac{2}{3}$ and $b = \frac{1}{3}$ thus the formula $\Sigma k^2 = \frac{1}{2} \cdot n \cdot (n+1) \cdot \left(\frac{2}{3} \cdot n + \frac{1}{3}\right)$.

For this method of finding suitable coefficients see Sect. 16.1.

Faulhaber developed the technique of comparing table columns to perfection; for example, he discovered the relationship

$n \cdot \Sigma k^2 - \Sigma k^3 = \Sigma(\Sigma k^2) - \Sigma k^2$ where $\Sigma(\Sigma k^2)$, the reduced form for $(1^2) + (1^2 + 2^2) + (1^2 + 2^2 + 3^2) + \ldots + (1^2 + 2^2 + 3^2 + \ldots + n^2)$ is noted.

| $n$ | $n^2$ | $\Sigma k^2$ | $\Sigma(\Sigma k^2)$ | $n \cdot \Sigma k^2$ | $k^3$ | $\Sigma k^3$ | $n \cdot \Sigma k^2 - \Sigma k^3$ | $\Sigma(\Sigma k^2) - \Sigma k^2$ |
|---|---|---|---|---|---|---|---|---|
| 1 | 1 | 1 | 1 | 1 | 1 | 1 | 0 | 0 |
| 2 | 4 | 5 | 6 | 10 | 8 | 9 | 1 | 1 |
| 3 | 9 | 14 | 20 | 42 | 27 | 36 | 6 | 6 |
| 4 | 16 | 30 | 50 | 120 | 64 | 100 | 20 | 20 |
| 5 | 25 | 55 | 105 | 275 | 125 | 225 | 50 | 50 |
| 6 | 36 | 91 | 196 | 546 | 216 | 441 | 105 | 105 |
| 7 | 49 | 140 | 334 | 980 | 343 | 784 | 196 | 196 |
| 8 | 64 | 204 | 540 | 1632 | 512 | 1296 | 336 | 336 |

This results in $\Sigma k^3 = (n+1) \cdot \Sigma k^2 - \Sigma(\Sigma k^2)$ so that one can also use it to make a direct calculation for $\Sigma(\Sigma k^2)$:

$\Sigma(\Sigma k^2) = (n+1) \cdot \Sigma k^2 - \Sigma k^3$.

For higher powers, Faulhaber tested various methods of decomposition until he found one that suited him. For example, he found that the sum $\Sigma k^4$ of the first $n$ fourth powers can be represented as the product of the form $(a \cdot \Sigma k + b) \cdot (\Sigma k^2)$ with $a = \frac{6}{5}$ and $b = -\frac{1}{5}$, see also, **A 16.1.**

Faulhaber's ingenious approaches cannot be further explored here; more details can be found in the following biography:

Ivo Schneider (1993): *Johannes Faulhaber: 1580–1635. Arithmetic master in a world of upheaval (Rechenmeister in einer Welt des Umbruchs)*. Birkhäuser, Basel.

In the following we will discuss further methods and impressive discoveries by other mathematicians.

**Suggestions for reflection and for investigations**
**A 16.1:** Add further columns to the above tables and check whether the following equations also apply:
$$\Sigma k^2 = (n+1) \cdot \Sigma k - \Sigma(\Sigma k) \text{ and } \Sigma k^4 = (n+1) \cdot \Sigma k^3 - \Sigma(\Sigma k^3).$$

## 16.1   Derivation of Sum Formulas using Arithmetic Sequences of Higher Order

Sums of powers of natural numbers can be determined with a simple approach, namely by means of arithmetic sequences of higher order.

---

**Definition**

**Arithmetic sequences of higher order**

A sequence of numbers $(a_n)_{n \in \mathbb{N}}$ where the corresponding sequence of differences $\Delta_1$ of two consecutive elements forms a constant sequence, is called as an **arithmetical sequence of the 1st order**.

An **arithmetical sequence of the $n$-th order** is a sequence of numbers where the corresponding sequence of differences $\Delta_n$ of two consecutive elements forms an arithmetic sequence of the $(n-1)$-th order.

Constant sequences can, therefore, also be called arithmetical sequences of the $0^{\text{th}}$ order. ◄

---

**Examples of arithmetical sequences**

- The sequence 0, 1, 2, 3, … of natural numbers is an arithmetical sequence of the 1st order, because its corresponding sequence of differences is the constant sequence 1, 1, 1, …
- The sequence 1, 3, 5, 7, … of odd natural numbers is also an arithmetical sequence of the 1st order, because its corresponding sequence of differences is the constant sequence 2, 2, 2, …
- The sequence 0, 1, 4, 9, 16, … of squares of natural numbers is an arithmetical sequence of the 2nd order, because its corresponding sequence of differences is 1, 3, 5, 7, …, the sequence of the odd natural numbers, thus an arithmetical sequence of the 1st order.
- The sequence of the triangular numbers 0, 1, 3, 6, 10, 15, … is an arithmetic sequence of the 2nd order, because its corresponding sequence of differences is 1, 2, 3, 4, 5, …, the sequence of natural numbers, that is, an arithmetic sequence of the 1st order.

The sequence of triangular numbers is formed by summing the natural numbers. Therefore, in the reverse process, that is, in calculating the difference of consecutive elements, the elements of the sequence of natural numbers result again.

- The sequence 0, 1, 8, 27, 64, 125, ... of the cubes of natural numbers is an arithmetic sequence of the 3rd order. Its corresponding sequence of differences has the elements 1, 7, 19, 37, 61, ..., whose corresponding sequence of differences has the elements 6, 12, 18, 24, .... This is an arithmetic sequence of the 1st order, because its corresponding sequence of differences is a constant sequence; thus the sequence 1, 7, 19, 37, 61, ... is an arithmetic sequence of the 2nd order and the sequence of cube numbers itself is an arithmetic sequence of the 3rd order (see the following table).

| $n$ | 0 | 1 | 2 | 3 | 4 | 5 | 6 | 7 | ... |
|---|---|---|---|---|---|---|---|---|---|
| $a_n$ | 0 | 1 | 8 | 27 | 64 | 125 | 216 | 343 | ... |
| $\Delta_1$ | 1 | 7 | 19 | 37 | 61 | 91 | 127 | ... | |
| $\Delta_2$ | 6 | 12 | 18 | 24 | 30 | 36 | ... | | |
| $\Delta_3$ | 6 | 6 | 6 | 6 | 6 | ... | | | |

- The sequence $s_n$ of the sums of squares of natural numbers, that is, 0, 1, 5, 14, ..., is an arithmetic sequence of the 3rd order, because we get the initial sequence of square numbers when we calculate the differences of consecutive elements, that is, an arithmetic sequence of the 2nd order (see the following table).

| $n$ | 0 | 1 | 2 | 3 | 4 | 5 | 6 | 7 | ... |
|---|---|---|---|---|---|---|---|---|---|
| $a_n$ | 0 | 1 | 4 | 9 | 16 | 25 | 36 | 49 | ... |
| $s_n$ | 0 | 1 | 5 | 14 | 30 | 55 | 91 | 140 | ... |
| $\Delta_1$ | 0 | 1 | 4 | 9 | 16 | 25 | 36 | 49 | ... |
| $\Delta_2$ | 1 | 3 | 5 | 7 | 9 | 11 | 13 | ... | ... |
| $\Delta_3$ | 2 | 2 | 2 | 2 | 2 | 2 | ... | | |

Of course, a calculation of the first elements of a sequence is not yet a proof for any statements about the order of an arithmetical sequence. To provide a formal proof, binomial formulas of higher order and some transformations are usually needed.

For example, for the sequence of cube numbers we get.

- $\Delta_1(n) = (n+1)^3 - n^3 = n^3 + 3 \cdot n^2 + 3 \cdot n + 1 - n^3 = 3 \cdot n^2 + 3 \cdot n + 1$
  so it's an arithmetic sequence of the 2nd order.
- $\Delta_2(n) = \left[3 \cdot (n+1)^2 + 3 \cdot (n+1) + 1\right] - \left(3 \cdot n^2 + 3 \cdot n + 1\right)$
  $= 3 \cdot n^2 + 6 \cdot n + 3 + 3 \cdot n + 3 + 1 - 3 \cdot n^2 - 3 \cdot n - 1 = 6 \cdot n + 6$

  ...which is an arithmetic sequence of the 1st order.
- $\Delta_3(n) = [6 \cdot (n+1) + 6] - (6 \cdot n + 6) = 6 \cdot n + 6 + 6 - 6 \cdot n - 6 = 6$,
  ...which is an arithmetical sequence of $0^{\text{th}}$ order.

Conversely, one can find for an arithmetic sequence of $0^{\text{th}}$ order, for example, $a_n = u$ with $u \in \mathbb{R}$, infinitely many arithmetical sequences $(b_n)_{n \in \mathbb{N}}$ of the 1st order, whose sequence of differences is equal to $a_n$, namely the linear sequences $b_n = u \cdot n + v$ with arbitrary $v \in \mathbb{R}$.

And also, for an arithmetic sequence of the 1st order, for example, $b_n = u \cdot n + v$ with $u, v \in \mathbb{R}$ an infinite number of arithmetic sequences $(c_n)_{n \in \mathbb{N}}$ of the 2nd order can be specified, whose sequence of differences is equal to $b_n$, namely the quadratic sequences $c_n = \frac{1}{2} \cdot u \cdot n^2 + \left(v - \frac{1}{2} \cdot u\right) \cdot n + w$ with arbitrary $w \in \mathbb{R}$.

**Suggestions for reflection and for investigations**

**A 16.2:** The quadratic sequence $(c_n)$ can be found by the following approach $c_n = \alpha \cdot n^2 + \beta \cdot n + \gamma$. With this approach you can determine the corresponding term for the sequence of differences and compare this with the equation $b_n = u \cdot n + v$. Thus you find a relationship between the coefficients $\alpha, \beta, \gamma$ and $u, v$, from which $u, v$ can be determined. Perform this calculation and compare the coefficients.

A calculation as in **A 16.2** is generally possible. That is, for arithmetic sequences of $k$-th order one can find an infinite number of arithmetic sequences of $(k+1)$-th order, whose sequence of differences is exactly the arithmetic sequence of $k$-th order one considers.

From the above examples and the information just given it is clear:

**Representation of the term of an arithmetic sequence using a polynomial function**

The elements of an arithmetical sequence of $k$-th order can be described by a polynomial function of $k$-th degree, that is, one has:
$$a(n) = a_k \cdot n^k + a_{k-1} \cdot n^{k-1} + \ldots + a_1 \cdot n^1 + a_0 \blacktriangleleft$$

In order to determine the formula for the sum $s(n)$ of an arithmetic sequence of $k$-th order, a polynomial function of $(k+1)$-th degree must be found. Then the first elements of the sum sequence can be used to determine the corresponding coefficients of the polynomial.

To derive a sum formula for the sequence of the $n$-th power of natural numbers, that is, an arithmetic sequence of $(n+1)$-th order, one needs $n+2$ equations. To be able to set up $n+2$ equations, it is sufficient to calculate the first $n+2$ elements of the sequence. If you start with 0, the system of equations is reduced to $n+1$ equations.

For the examples considered in the following, you need the first elements of the sum sequence of natural numbers, square numbers, cube numbers, and the fourth power of natural numbers. Here, one can refer back to the table in Fig. 16.1. The sum symbol $\Sigma$ (= Greek capital letter Sigma, corresponds to the letter S for sum) introduced by Gottfried Wilhelm Leibniz is used as symbol for the calculated sums of the mentioned numbers.

| Powers | | | | Sum of powers | | | |
|---|---|---|---|---|---|---|---|
| $k$ | $k^2$ | $k^3$ | $k^4$ | $\Sigma k$ | $\Sigma k^2$ | $\Sigma k^3$ | $\Sigma k^4$ |
| 0 | 0 | 0 | 0 | 0 | 0 | 0 | 0 |
| 1 | 1 | 1 | 1 | 1 | 1 | 1 | 1 |
| 2 | 4 | 8 | 16 | 3 | 5 | 9 | 17 |
| 3 | 9 | 27 | 81 | 6 | 14 | 36 | 98 |
| 4 | 16 | 64 | 256 | 10 | 30 | 100 | 354 |
| 5 | 25 | 125 | 625 | 15 | 55 | 225 | 979 |
| 6 | 36 | 216 | 1296 | 21 | 91 | 441 | 2275 |
| 7 | 49 | 343 | 2401 | 28 | 140 | 784 | 4676 |
| 8 | 64 | 512 | 4096 | 36 | 204 | 1296 | 8772 |
| 9 | 81 | 729 | 6561 | 45 | 285 | 2025 | 15333 |
| 10 | 100 | 1000 | 10000 | 55 | 385 | 3025 | 25333 |

**Fig. 16.1** Natural numbers and their powers (up to 4th degree) and their sums

**Example 1: Calculating the sum of the first n square numbers**

Start with a polynomial function of 3rd degree:
$$s(n) = a \cdot n^3 + b \cdot n^2 + c \cdot n + d$$
Figure 16.1 shows that:
$$s(0) = 0; \quad s(1) = 1; \quad s(2) = 5; \quad s(3) = 14$$

Because of $s(0) = 0, d = 0$ follows immediately.

The following linear system of equations with three equations and three variables must be solved:
$$\begin{vmatrix} 1a + 1b + 1c = 1 \\ 8a + 4b + 2c = 5 \\ 27a + 9b + 3c = 14 \end{vmatrix}$$

This has the solution $\left( \frac{1}{3}, \frac{1}{2}, \frac{1}{6} \right)$.

The steps which are necessary to determine the solution are explained below.

We therefore get: $1^2 + 2^2 + 3^2 + \ldots + n^2 = \frac{1}{3}n^3 + \frac{1}{2}n^2 + \frac{1}{6}n$

If the term is factored into linear factors, the formula is easier to remember:

$$\frac{1}{3}n^3 + \frac{1}{2}n^2 + \frac{1}{6}n = \frac{1}{3}n \cdot \left( n^2 + \frac{3}{2}n + \frac{1}{2} \right) = \frac{1}{3} \cdot n \cdot (n+1) \cdot \left( n + \frac{1}{2} \right) = \frac{1}{6} \cdot n \cdot (n+1) \cdot (2n+1)$$

This formula was already derived in Chap. 2 (see there formula (2.4)).

**Example 2: Calculating the sum of the first n cube numbers**

Start with a polynomial function of 4th degree:
$$s(n) = a \cdot n^4 + b \cdot n^3 + c \cdot n^2 + d \cdot n + e$$
Figure 16.1 shows that
$$s(0) = 0; \quad s(1) = 1; \quad s(2) = 9; \quad s(3) = 36; \quad s(4) = 100$$
Because of $s(0) = 0, e = 0$ follows immediately.

The following linear system of equations with four equations and four variables must be solved:
$$\begin{vmatrix} 1a + 1b + 1c + 1d = 1 \\ 16a + 8b + 4c + 2d = 9 \\ 81a + 27b + 9c + 3d = 36 \\ 256a + 64b + 16c + 4d = 100 \end{vmatrix}$$

This has the solution $\left( \frac{1}{4}, \frac{1}{2}, \frac{1}{4}, 0 \right)$.

We therefore get

$$1^3 + 2^3 + 3^3 + \ldots + n^3 = \frac{1}{4}n^4 + \frac{1}{2}n^3 + \frac{1}{4}n^2 = \frac{1}{4} \cdot n^2 \cdot (n^2 + 2n + 1) = \left[ \frac{1}{2} \cdot n \cdot (n+1) \right]^2,$$

see also formula (2.5).

**Solving linear equation systems by elementary row operations**

When solving systems of linear equations, a systematic procedure is recommended. In this method, one of the rows (equations) is multiplied by a suitable number and added to another line – with the aim of eliminating a variable in the other line.

**Application of the addition method for example 1**

$$\begin{vmatrix} 1a + 1b + 1c = 1 \\ 8a + 4b + 2c = 5 \\ 27a + 9b + 3c = 14 \end{vmatrix} \Leftrightarrow \begin{vmatrix} 1a + 1b + 1c = 1 \\ 7a + 3b + 1c = 4 \\ 19a + 5b + 1c = 9 \end{vmatrix} \Leftrightarrow \begin{vmatrix} 1a + 1b + 1c = 1 \\ 6a + 2b = 3 \\ 12a + 2b = 5 \end{vmatrix}$$

$$\Leftrightarrow \begin{vmatrix} 1a + 1b + 1c = 1 \\ 6a + 2b = 3 \\ 6a = 2 \end{vmatrix} \Leftrightarrow \begin{vmatrix} c = \frac{1}{6} \\ b = \frac{1}{2} \\ a = \frac{1}{3} \end{vmatrix}$$

Hold the 1st line, add the $(-1)$-multiple of the 1st row to the 2nd row, the $(-1)$-multiple of the 2nd row to the 3rd row. If you repeat this, you have already eliminated the last variable. Then, in addition to the 1st row, you also hold the 2nd row and add the $(-1)$-multiple of the 2nd row to the 3rd row, so that the second variable is eliminated there. The $3 \times 3$ equation system has now a triangular shape and one can determine the variables step by step from bottom to top.

That this works so wonderfully is due to the coefficients: The coefficients of the variables $a$, $b$, $c$ form arithmetic sequences of 3rd, 2nd, and 1st order respectively: adding the $(-1)$-multiple of the previous line corresponds to the determination of the sequence of differences.

**Application of the addition method for example 2**

$$\begin{vmatrix} 1a + 1b + 1c + 1d = 1 \\ 16a + 8b + 4c + 2d = 9 \\ 81a + 27b + 9c + 3d = 36 \\ 256a + 64b + 16c + 4d = 100 \end{vmatrix} \Leftrightarrow \begin{vmatrix} 1a + 1b + 1c + 1d = 1 \\ 15a + 7b + 3c + 1d = 8 \\ 65a + 19b + 5c + 1d = 27 \\ 175a + 37b + 7c + 1d = 64 \end{vmatrix} \Leftrightarrow \begin{vmatrix} 1a + 1b + 1c + 1d = 1 \\ 14a + 6b + 2c = 7 \\ 50a + 12b + 2c = 19 \\ 110a + 18b + 2c = 37 \end{vmatrix}$$

$$\Leftrightarrow \begin{vmatrix} 1a + 1b + 1c + 1d = 1 \\ 14a + 6b + 2c = 7 \\ 36a + 6b = 12 \\ 60a + 6b = 18 \end{vmatrix} \Leftrightarrow \begin{vmatrix} 1a + 1b + 1c + 1d = 1 \\ 14a + 6b + 2c = 7 \\ 36a + 6b = 12 \\ 24a = 6 \end{vmatrix} \Leftrightarrow \begin{vmatrix} d = 0 \\ c = \frac{1}{4} \\ b = \frac{1}{2} \\ a = \frac{1}{4} \end{vmatrix}$$

**Suggestions for reflection and for investigations**

**A 16.3:** Set up the linear system of equations required to determine the sum of the fourth [fifth] powers of the natural numbers. Solve the linear system of equations and thus determine the coefficients of the polynomial function $s(n)$, with which the sum of the fourth [fifth] powers can be calculated directly.

## 16.2   Determination of Coefficients by Comparing Consecutive Elements in the Sum Sequence

The formula for the sum of the powers of the $(n+1)$ first natural numbers can also be determined recursively (backwards): one obtains the sum $s_k(n+1)$ of the $k$-th powers of the first $(n+1)$ natural numbers, by adding the k-th power of the natural number $(n+1)$ to the sum $s_k(n)$ of the $k$-th power of the first $n$ natural numbers:

$$s_k(n+1) = s_k(n) + (n+1)^k$$

**Example: Recursive determination of coefficients**

The sum of the first $n+1$ square numbers is calculated recursively from $s_2(n+1) = s_2(n) + (n+1)^2$.

The polynomial approach results in

$$s_2(n) = a \cdot n^3 + b \cdot n^2 + c \cdot n + d$$

as well as

$$s_2(n+1) = a \cdot (n+1)^3 + b \cdot (n+1)^2 + c \cdot (n+1) + d.$$

Therefore, we get the following:

$$\left[a \cdot n^3 + b \cdot n^2 + c \cdot n + d\right] + (n+1)^2 = a \cdot (n+1)^3 + b \cdot (n+1)^2 + c \cdot (n+1) + d$$

The binomial terms can be developed:

$$a \cdot n^3 + b \cdot n^2 + c \cdot n + d + \left(n^2 + 2n + 1\right)$$
$$= a \cdot \left(n^3 + 3n^2 + 3n + 1\right) + b \cdot \left(n^2 + 2n + 1\right) + c \cdot (n+1) + d$$

Ordered by powers of $n$:

$$(a - a) \cdot n^3 + (b + 1 - 3a - b) \cdot n^2 + (c + 2 - 3a - 2b - c) \cdot n + (d + 1 - a - b - c - d) = 0,$$

so $(1 - 3a) \cdot n^2 + (2 - 3a - 2b) \cdot n + (1 - a - b - c) = 0$.

Since this equation applies to any natural number $n \in \mathbb{N}$, the terms in brackets must be zero.

This is how you get the coefficients step-by-step: from $1 - 3a = 0$ you get the $a = \frac{1}{3}$, from $2 - 3a - 2b = 0$ you get $b = \frac{1}{2}$, and from $1 - a - b - c = 0$ you get $c = \frac{1}{6}$. From $s(0) = 0$ it results $d = 0$.

In principle, following this method of comparing the coefficients, one can determine the coefficients of the polynomial functions of $(n+1)$-th degree, and so obtain a formula for the sum of $n$-th powers of natural numbers. For this the general binomial theorem is needed.

---

**Formula**

**General binomial theorem (powers of a sum of two real numbers)**

For $a, b \in \mathbb{R}$ we get:

$$(a+b)^2 = 1a^2b^0 + 2a^1b^1 + 1a^0b^2$$
$$(a+b)^3 = 1a^3b^0 + 3a^2b^1 + 3a^1b^2 + 1a^0b^3$$
$$(a+b)^4 = 1a^4b^0 + 4a^3b^1 + 6a^2b^2 + 4a^1b^3 + 1a^0b^4$$

and so on ◀

The factors of the powers of $a$ and $b$ are – because of the connection with the binomial formulas – called **binomial coefficients**. They can be seen, for example, in Pascal's triangle (see the following scheme and the stamp issued by Korea (South) on the occasion of the International Congress of Mathematicians (ICM) in Seoul in 2014).

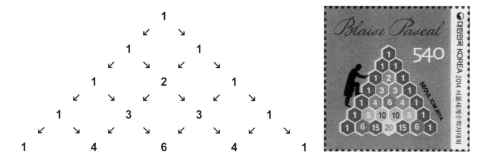

Binomial coefficients can also be calculated directly using a formula with a quotient, where $k$ factors are in the numerator and in the denominator. However, this is not really a fraction, as all factors of the denominator can be cancelled.

---

**Formula**

**Calculation of binomial coefficients**

For $n, k \in \mathbb{N}$ and $0 \leq k \leq n$ we get: $\binom{n}{k} = \frac{n \cdot (n-1) \cdot (n-2) \cdot \ldots \cdot (n-k+1)}{1 \cdot 2 \cdot 3 \cdot \ldots \cdot k}$. ◀

Using the notation $\binom{n}{k}$ for the binomial coefficients, the binomial formulas can be noted as follows

---

**Formula**

**General binomial theorem (power of a sum of two real numbers)**

For $a, b \in \mathbb{R}$ we have:

$$(a+b)^2 = \binom{2}{0} \cdot a^2 b^0 + \binom{2}{1} \cdot a^1 b^1 + \binom{2}{2} \cdot a^0 b^2$$

$$(a+b)^3 = \binom{3}{0} \cdot a^3 b^0 + \binom{3}{1} \cdot a^2 b^1 + \binom{3}{2} \cdot a^1 b^2 + \binom{3}{3} \cdot a^0 b^3$$

$$(a+b)^4 = \binom{4}{0} \cdot a^4 b^0 + \binom{4}{1} \cdot a^3 b^1 + \binom{4}{2} \cdot a^2 b^2 + \binom{4}{3} \cdot a^1 b^3 + \binom{4}{4} \cdot a^0 b^4$$

and so on ◀

---

**Suggestions for reflection and for investigations**

**A 16.4:** Determine the formula for calculating the sum of the first $n$ cubic numbers (the sum of the first fourth powers) according to the recursive method.

**A 16.5:** In the example above, the coefficients $a$, $b$, $c$, $d$ were deduced from the equation:

$$a \cdot n^3 + b \cdot n^2 + c \cdot n + d + (n^2 + 2n + 1)$$
$$= a \cdot (n^3 + 3n^2 + 3n + 1) + b \cdot (n^2 + 2n + 1) + c \cdot (n+1) + d$$

In this equation, replace the binomial coefficients by the symbols introduced above. Write down the equations in the last step of the transformations.

---

## 16.3  Alhazen's Derivation of the Sum Formulas for Higher Powers

In Sect. 2.5 the approach Abu Ali al-Hasan ibn al-Haitham (Alhazen) used to derive a formula for the sum of the first $n$ squares was explained. Square numbers are represented by squares in the rectangular figure used for this purpose.

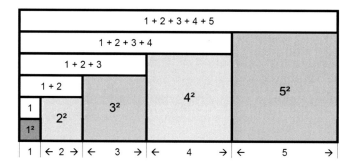

Ibn al-Haitham also found a way to derive a formula for the sums of higher powers. In principle he used the same rectangular figure, but interpreted it differently, namely that the squares should represent rectangles with heights 1, 2, 3, ... and "widths" $1^2$, $2^2$, $3^2$..., that is, the "squares" have the area $1^3$, $2^3$, $3^3$, ...

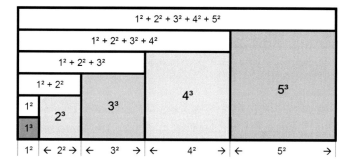

Thus, in the modified rectangular figure, the following relationship therefore holds:

The sum of the "areas" of the individual subareas is equal to the area of the rectangle with the width $1^2 + 2^2 + 3^2 + \ldots + n^2$ and the height $n+1$:

$$\left[1^3 + 2^3 + 3^3 + \ldots + n^3\right] + \left[1^2 + \left(1^2 + 2^2\right) + \left(1^2 + 2^2 + 3^2\right) + \ldots + \left(1^2 + 2^2 + 3^2 + \ldots + n^2\right)\right]$$
$$= (n+1) \cdot \left(1^2 + 2^2 + 3^2 + \ldots + n^2\right)$$

After applying the sum formulas

$$1 + 2 + 3 + \ldots + n = \tfrac{1}{2}n^2 + \tfrac{1}{2}n \text{ and } 1^2 + 2^2 + 3^2 + \ldots + n^2 = \tfrac{1}{3}n^3 + \tfrac{1}{2}n^2 + \tfrac{1}{6}n.$$

which have been derived before, it follows:

$$\left[1^3 + 2^3 + \ldots + n^3\right] + \left[\left(\tfrac{1}{3} \cdot 1^3 + \tfrac{1}{2} \cdot 1^2 + \tfrac{1}{6} \cdot 1\right) + \left(\tfrac{1}{3} \cdot 2^3 + \tfrac{1}{2} \cdot 2^2 + \tfrac{1}{6} \cdot 2\right) + \ldots + \left(\tfrac{1}{3} \cdot n^3 + \tfrac{1}{2} \cdot n^2 + \tfrac{1}{6} \cdot n\right)\right]$$
$$= (n+1) \cdot \left(1^2 + 2^2 + 3^2 + \ldots + n^2\right)$$

And after rearranging and combining terms of the same kind:

$$\left[1^3 + 2^3 + \ldots + n^3\right] + \left[\tfrac{1}{3} \cdot \left(1^3 + 2^3 + \ldots + n^3\right) + \tfrac{1}{2} \cdot \left(1^2 + 2^2 + \ldots + n^2\right) + \tfrac{1}{6} \cdot \left(1 + 2 + \ldots + n\right)\right]$$
$$= (n+1) \cdot \left(1^2 + 2^2 + 3^2 + \ldots + n^2\right)$$
$$\Leftrightarrow \tfrac{4}{3} \cdot \left(1^3 + 2^3 + \ldots + n^3\right) = \left(n + \tfrac{1}{2}\right) \cdot \left(1^2 + 2^2 + 3^2 + \ldots + n^2\right) - \tfrac{1}{6} \cdot \left(1 + 2 + \ldots + n\right)$$
$$\Leftrightarrow 1^3 + 2^3 + \ldots + n^3 = \tfrac{3}{4} \cdot \left(n + \tfrac{1}{2}\right) \cdot \left(1^2 + 2^2 + 3^2 + \ldots + n^2\right) - \tfrac{3}{4} \cdot \tfrac{1}{6} \cdot \left(1 + 2 + \ldots + n\right)$$

If we now use the sum formulas from above, we obtain

$$1^3 + 2^3 + \ldots + n^3 = \left(\tfrac{3}{4}n + \tfrac{3}{8}\right) \cdot \left(\tfrac{1}{3}n^3 + \tfrac{1}{2}n^2 + \tfrac{1}{6}n\right) - \tfrac{1}{8} \cdot \left(\tfrac{1}{2}n^2 + \tfrac{1}{2}n\right),$$

so

$$1^3 + 2^3 + \ldots + n^3 = \tfrac{1}{4}n^4 + \tfrac{3}{8}n^3 + \tfrac{1}{8}n^2 + \tfrac{1}{8}n^3 + \tfrac{3}{16}n^2 + \tfrac{1}{16}n - \tfrac{1}{16}n^2 - \tfrac{1}{16}n,$$

and further

$$1^3 + 2^3 + \ldots + n^3 = \tfrac{1}{4}n^4 + \tfrac{1}{2}n^3 + \tfrac{1}{4}n^2.$$

This term can be further transformed to:

$$1^3 + 2^3 + \ldots + n^3 = \tfrac{1}{4} \cdot n^2 \cdot (n+1)^2 = \left(\tfrac{1}{2} \cdot n \cdot (n+1)\right)^2$$

This idea is generally applicable and leads to sum formulas for any powers.

> **Suggestions for reflection and for investigations**
> **A 16.6:** Derive a formula for the sum of the fourth powers of the first $n$ natural numbers using the method of al-Haitham. Make a sketch for this. Use the sum formulas for natural numbers, square numbers, and cube numbers for the transformations.

For the special case of the formula for the sum of squares George Pólya, an American mathematician of Hungarian origin, suggests in volume 1 of *Mathematics and plausible reasoning* (Princeton University Press) comparing the sum sequences of the sequence of natural numbers and square numbers by looking at their quotients (see Fig. 16.2).

Obviously all the quotients of the considered elements of the two sum sequences ($\sum k$ and $\sum k^2$) are equal to $(2n+1)/3$.

| $k$ | $k^2$ | $\sum k$ | $\sum k^2$ | $\sum k^2 / \sum k$ |
|---|---|---|---|---|
| 1 | 1 | 1 | 1 | 1/1 = 3/3 |
| 2 | 4 | 3 | 5 | 5/3 |
| 3 | 9 | 6 | 14 | 14/6 = 7/3 |
| 4 | 16 | 10 | 30 | 30/10 = 9/3 |
| 5 | 25 | 15 | 55 | 55/15 = 11/3 |
| 6 | 36 | 21 | 91 | 91/21 = 13/3 |
| 7 | 49 | 28 | 140 | 140/28 = 15/3 |
| 8 | 64 | 36 | 204 | 204/36 = 17/3 |
| 9 | 81 | 45 | 285 | 285/45 = 19/3 |
| 10 | 100 | 55 | 385 | 385/55 = 21/3 |

**Fig. 16.2** Comparison of the sums of natural numbers and square numbers

From this one can deduce a formula for the sequence of sums of square numbers.

$$\frac{\sum\limits_{k=1}^{n} k^2}{\sum\limits_{k=1}^{n} k} = \frac{\sum\limits_{k=1}^{n} k^2}{\frac{1}{2} \cdot n \cdot (n+1)} = \frac{2n+1}{3}$$

and thus $\sum\limits_{k=1}^{n} k^2 = \frac{1}{2} \cdot n \cdot (n+1) \cdot \frac{2n+1}{3} = \frac{1}{6} \cdot n \cdot (n+1) \cdot (2n+1)$.

As this is only a method to *discover* the formula but does not *prove* it, Pólya subsequently completes the proof by induction.

This nice way of *discovering* a sum formula, however, only succeeds in the case $\sum k^2/\sum k$ as you can see from al-Haitham's derivation (see also, **A 16.7**).

---

**Suggestions for reflection and for investigations**

**A 16.7:** In the transformations resulting from al-Haitham's approach, one comes to a point where knowledge of the sums of lower powers is needed.

In deriving the sum of the square numbers, this is the equation:

$$1^2 + 2^2 + 3^2 + \ldots + n^2 = \left(\tfrac{2}{3}n + \tfrac{1}{3}\right) \cdot (1 + 2 + 3 + \ldots + n)$$

in deriving the sum of the cube numbers, the equation is:

$$1^3 + 2^3 + \ldots + n^3 = \tfrac{3}{4} \cdot \left(n + \tfrac{1}{2}\right) \cdot \left(1^2 + 2^2 + 3^2 + \ldots + n^2\right) - \tfrac{3}{4} \cdot \tfrac{1}{6} \cdot (1 + 2 + \ldots + n)$$

Explain why it can be seen at this point that Pólya's idea cannot be applied to the sum formula for cubic numbers.

---

## 16.4 Thomas Harriot Discovers a Connection between Triangular and Tetrahedral Numbers

Since ancient times, mathematicians have been dealing with **figurate numbers,** that is, with numbers that can be represented by a particular geometric figure.

The **polygonal numbers** are based on the geometric figure of a regular polygon.

These include, for example, the **triangular numbers** (1, 3, 6, 10, 15, 21, …),

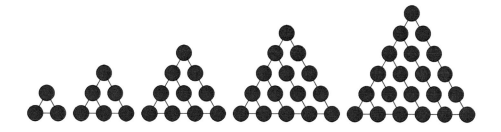

the square numbers $(1, 4, 9, 16, 25, 36, \ldots)$.

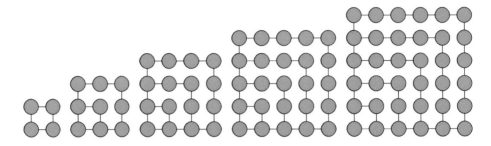

and the pentagonal numbers $(1, 5, 12, 22, 35, 51, \ldots)$:

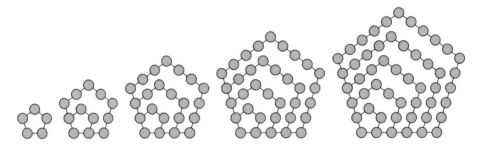

(In the figures above, the first elements of the sequences consisting of only one "dot" has been omitted in each case).

A sequence of polygonal numbers forms an arithmetic sequence of 2nd order (since the increase of the elements is based on an arithmetic sequence of the 1st order).

The sum sequence belonging to a sequence of polygonal numbers is called a sequence of **pyramidal numbers**. If you look at three-dimensional balls instead of two-dimensional circles, they can be three-dimensionally "stacked" to form pyramids.

From the sequence of the triangular numbers $1, 3, 6, 10, 15, 21, \ldots$ you get by summation the sequence of the **tetrahedral numbers** $1, 4, 10, 20, 35, 56, \ldots$ - the following figure illustrates the tetrahedral number 35:

From the sequence of the square numbers 1, 4, 9, 16, 25, 36, … you get by summation
the sequence of the **square pyramidal numbers** 1, 5, 14, 30, 55, 91, … – the following
figure illustrates the square pyramidal number 30:

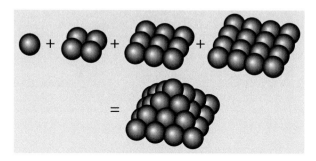

(Source: Square pyramidal number, David Eppstein/English Wikipedia/public domain)

The English mathematician Thomas Harriot (1560–1621) made a remarkable discovery when studying triangular and tetrahedral numbers. In his paper *De Numeris
Triangularibus et inde De Progressionibus Arithmeticis Magisteria Magna* ("On
Triangular Numbers and a Great Theorem on Arithmetic Progressions", around 1611) he
set up a table with the side lengths *(laterals)*, the triangular numbers *(triangulars)*, and
the pyramidal numbers *(pyramidals)*, and added a row and column for the length unit
*(units)*.

He discovered the following: If you add up all the numbers in a row (or column) up
to a certain element, the sum is exactly equal to the number that is in this position in the
next row (or column). Conversely, the numbers from the previous row (or column) are
obtained by subtracting the elements of a row (or column).

If this property is taken into consideration, the original table can be extended with
additional rows and columns as follows

| | | | | | | | |
|---|---|---|---|---|---|---|---|
| units | 1 | 1 | 1 | 1 | 1 | 1 | 1 |
| laterals | 1 | 2 | 3 | 4 | 5 | 6 | 7 |
| triangulars | 1 | 3 | 6 | 10 | 15 | 21 | 28 |
| pyramidals | 1 | 4 | 10 | 20 | 35 | 56 | 84 |
| | 1 | 5 | 15 | 35 | 70 | 126 | 210 |
| | 1 | 6 | 21 | 56 | 126 | 252 | 462 |
| | 1 | 7 | 28 | 84 | 210 | 462 | 924 |

Harriot then applied this idea to derive arithmetic sequences of higher order.

| 2 | 5 | 4 | 1 | | | | |
|---|---|---|---|---|---|---|---|
| | 7 | 9 | 5 | a | b | c | d |
| 2 | 9 | 16 | 14 | a | b+a | c+b | d+c |
| 2 | 11 | 25 | 30 | a | b+2a | c+2b+a | d+2c+b |
| 2 | | 36 | 55 | a | b+3a | c+3b+3a | d+3c+3b+a |
| | | | 91 | | | c+4b+6a | d+4c+6b+4a |
| | | | | | | | d+5c+10b+10a |

In the table on the left, the last column shows the elements of the sum sequence of the square numbers of the first $n$ natural numbers: 1, 5, 14, 30, 55, 91, .... The preceding columns contain the elements of the corresponding sequence of differences. The entire table is clearly defined by the initial values 2, 5, 4, and 1 in the column headers highlighted in green.

In the table on the right, you can generally see which next elements result from the initial values $a$, $b$, $c$, and $d$. If you look at the last column in the table on the right, you will notice the following:

The elements of this sequence of numbers can be calculated as the sum of multiples of the four initial values, and the numbers from Pascal's triangle, that is, the binomial coefficients, appear as factors of the initial values. In contrast to the binomial formulas, however, the sums here consist of not more than four summands.

The last column in the table on the left contains the first six elements of the sum sequence of the square numbers. As can be seen from the table on the right by comparison, the 6th element, for example, is calculated as the sum of products of the initial numbersand binomial coefficients:

$$1 \cdot d + 5 \cdot c + 10 \cdot b + 10 \cdot a = 1 \cdot 1 + 5 \cdot 4 + 10 \cdot 5 + 10 \cdot 2 = 91$$

In order to calculate the 6th element of the sum sequence, the first four binomial coefficients from the 5th row of Pascal's triangle are needed.

In order to generally calculate the $n$-th element of the sum sequence of the square numbers, one needs the first four binomial coefficients from the $(n-1)$-th row of Pascal's triangle.

$$1^2 + 2^2 + 3^2 + \ldots + n^2 = 1 \cdot \binom{n-1}{0} + 4 \cdot \binom{n-1}{1} + 5 \cdot \binom{n-1}{2} + 2 \cdot \binom{n-1}{3}$$

$$= 1 + 4 \cdot \frac{n-1}{1} + 5 \cdot \frac{(n-1) \cdot (n-2)}{2 \cdot 1} + 2 \cdot \frac{(n-1) \cdot (n-2) \cdot (n-3)}{3 \cdot 2 \cdot 1}$$

$$= \ldots = \tfrac{1}{3}n^3 + \tfrac{1}{2}n^2 + \tfrac{1}{6}n$$

$$= \tfrac{1}{6} \cdot n \cdot (n+1) \cdot (2n+1)$$

**Suggestions for reflection and for investigations**
**A 16.8:** Justify: If Harriot had started the sum sequence with the element 0, then he would have had to calculate the $6^{th}$ element 91 as follows:

$$\mathbf{1 \cdot d + 6 \cdot c + 15 \cdot b + 20 \cdot a = 1 \cdot 0 + 6 \cdot 1 + 15 \cdot 3 + 20 \cdot 2 = 91}$$

Derive the sum formula for square numbers from this, too.

Harriot's method can also be applied accordingly, if a formula for the sum of the first $n$ cube numbers has to be determined.

| 6 | 12 | 7 | 1 | 0 | | | | | |
|---|---|---|---|---|---|---|---|---|---|

Table (left):

| | | | | | **0** |
|---|---|---|---|---|---|
| | | | | **1** | 1 |
| | | | **7** | 8 | 9 |
| | | **12** | 19 | 27 | 36 |
| **6** | 18 | 37 | 64 | 100 | |
| 6 | 24 | 61 | 125 | 225 | |
| 6 | 30 | 91 | 216 | | |
| | | | 441 | | |

Table (right):

| | | | **d** | **e** |
|---|---|---|---|---|
| | | **c** | d+c | e+d |
| | **b** | c+b | d+2c+b | e+2d+c |
| **a** | b+a | c+2b+a | d+3c+3b+a | e+3d+3c+b |
| a | b+2a | c+3b+3a | d+4c+6b+4a | e+4d+6c+4b+a |
| a | b+3a | c+4b+6a | d+5c+10b+10a | e+5d+10c+10b+5a |
| | | | | e+6d+15c+20b+15a |

The last column of the table on the left contains the first elements of the sum sequence, starting with the initial value 0 (therefore the penultimate column contains the elements of the sequence of cube numbers). The terms in the last column of the table on the right consist of not more than five summands.

For the $n$-th element of the sum sequence results here:

$$1^3 + 2^3 + 3^3 + \ldots + n^3 = 0 \cdot \binom{n}{0} + 1 \cdot \binom{n}{1} + 7 \cdot \binom{n}{2} + 12 \cdot \binom{n}{3} + 6 \cdot \binom{n}{4}$$

$$= 0 + 1 \cdot \frac{n}{1} + 7 \cdot \frac{n \cdot (n-1)}{2 \cdot 1} + 12 \cdot \frac{n \cdot (n-1) \cdot (n-2)}{3 \cdot 2 \cdot 1}$$

$$+ 6 \cdot \frac{n \cdot (n-1) \cdot (n-2) \cdot (n-3)}{4 \cdot 3 \cdot 2 \cdot 1}$$

$$= \ldots = \frac{1}{4}n^4 + \frac{1}{2}n^3 + \frac{1}{4}n^2$$

**Suggestions for reflection and for investigations**
**A 16.9:** Determine a formula for the sum of the fourth powers using Harriot's method.

## 16.5   Fermat's Discovery

The Frenchman Pierre de Fermat (1607–1665) worked as a lawyer in Toulouse; in his free time he was intensively engaged in mathematical problems. He was particularly fond of number theory.

In 1994, when Fermat's Last Theorem (FLT) was finally proved, namely that for $n>2$, the $n$-th power of a natural number cannot be represented as the sum of $n$-th powers of two natural numbers (not equal to zero), this was worth a stamp for the French postal service. However, the text of the stamp still contains the birth year 1601, which later turned out to be incorrect.

Fermat had an extensive correspondence with various mathematicians about his discoveries, including Marin Mersenne in Paris, who in turn informed numerous correspondents throughout Europe about the reports he received.

One of Fermat's discoveries was the fact that triangular and tetrahedral numbers can be written as fractional terms with ascending factors. (This is surprising because these numbers are *quantities* that is, natural numbers. The fractional terms are actually not real fractions, because all factors in the denominator can be canceled).

The second and third column of the table in Fig. 16.3 show the fractional terms he found, with which the triangular and tetrahedral numbers can be calculated. The table also contains two further columns in which products with four or five ascending factors are listed (in the numerator beginning with $n$, in the denominator beginning with 1).

The following striking property applies not only to the first two columns of Fig. 16.3:

If one determines the sum of the first $n$ numbers in a column, one obtains the number which is in the $n$-th position in the next column.

For example, the sum of the first five numbers of the 2nd column, that is $1 + 3 + 6 + 10 + 15 = 35$, is in the 5$^{th}$ position of the 3rd column (highlighted in yellow), and the sum of the first four numbers of the 3rd column, that is, $1 + 5 + 15 + 35 = 56$ is placed in the 4$^{th}$ position of the 4$^{th}$ column (highlighted in green).

Since the summands of the 2nd column can generally be expressed in the form $\frac{n\cdot(n+1)}{1\cdot 2}$, this means $\frac{1\cdot 2}{1\cdot 2} + \frac{2\cdot 3}{1\cdot 2} + \frac{3\cdot 4}{1\cdot 2} + \ldots + \frac{n\cdot(n+1)}{1\cdot 2} = \frac{n\cdot(n+1)\cdot(n+2)}{1\cdot 2\cdot 3}$.

| n | $\dfrac{n \cdot (n+1)}{1 \cdot 2}$ | $\dfrac{n \cdot (n+1) \cdot (n+2)}{1 \cdot 2 \cdot 3}$ | $\dfrac{n \cdot (n+1) \cdot (n+2) \cdot (n+3)}{1 \cdot 2 \cdot 3 \cdot 4}$ | $\dfrac{n \cdot (n+1) \cdot (n+2) \cdot (n+3) \cdot (n+4)}{1 \cdot 2 \cdot 3 \cdot 4 \cdot 5}$ |
|---|---|---|---|---|
| 1 | 1 | 1 | 1 | 1 |
| 2 | 3 | 4 | 5 | 6 |
| 3 | 6 | 10 | 15 | 21 |
| 4 | 10 | 20 | 35 | 56 |
| 5 | 15 | 35 | 70 | 126 |
| 6 | 21 | 56 | 126 | 252 |
| 7 | 28 | 84 | 210 | 462 |
| 8 | 36 | 120 | 330 | 792 |
| 9 | 45 | 165 | 495 | 1287 |
| 10 | 55 | 220 | 715 | 2002 |

**Fig. 16.3**  Fermat's table with fractional terms

and accordingly for the numbers in the 3rd column

$$\frac{1 \cdot 2 \cdot 3}{1 \cdot 2 \cdot 3} + \frac{2 \cdot 3 \cdot 4}{1 \cdot 2 \cdot 3} + \frac{3 \cdot 4 \cdot 5}{1 \cdot 2 \cdot 3} + \ldots + \frac{n \cdot (n+1) \cdot (n+2)}{1 \cdot 2 \cdot 3} = \frac{n \cdot (n+1) \cdot (n+2) \cdot (n+3)}{1 \cdot 2 \cdot 3 \cdot 4}$$

and so on.

One can now multiply these equations by the denominator on the left and one gets interesting formulas:

---

**Formulas**

**Calculation of the sum of products of consecutive natural numbers**

$$1 \cdot 2 + 2 \cdot 3 + 3 \cdot 4 + \ldots + n \cdot (n+1) = \frac{n \cdot (n+1) \cdot (n+2)}{3}$$

$$1 \cdot 2 \cdot 3 + 2 \cdot 3 \cdot 4 + 3 \cdot 4 \cdot 5 + \ldots + n \cdot (n+1) \cdot (n+2) = \frac{n \cdot (n+1) \cdot (n+2) \cdot (n+3)}{4}$$

$$1 \cdot 2 \cdot 3 \cdot 4 + 2 \cdot 3 \cdot 4 \cdot 5 + 3 \cdot 4 \cdot 5 \cdot 6 + \ldots + n \cdot (n+1) \cdot (n+2) \cdot (n+3)$$
$$= \frac{n \cdot (n+1) \cdot (n+2) \cdot (n+3) \cdot (n+4)}{5}$$

and so on. ◄

---

The left side of the first of these two equations can also be noted as follows:

$$1 \cdot (1+1) + 2 \cdot (2+1) + 3 \cdot (3+1) + \ldots + n \cdot (n+1)$$
$$= \left(1^2 + 1\right) + \left(2^2 + 2\right) + \left(3^2 + 3\right) + \ldots + \left(n^2 + n\right)$$
$$= \left(1^2 + 2^2 + 3^2 + \ldots + n^2\right) + (1 + 2 + 3 + \ldots + n)$$
$$= (1^2 + 2^2 + 3^2 + \ldots + n^2) + \tfrac{1}{2} \cdot n \cdot (n+1),$$

so:

$$1^2 + 2^2 + 3^2 + \ldots + n^2 = \frac{n \cdot (n+1) \cdot (n+2)}{3} - \frac{n \cdot (n+1)}{2}$$

$$= \frac{n \cdot (n+1)}{6} \cdot [2 \cdot (n+2) - 3]$$

$$= \frac{n \cdot (n+1) \cdot (2n+1)}{6}$$

Using the same stepwise procedure, one can thus derive sum formulas for higher powers of natural numbers.

**Suggestions for reflection and for investigations**
**A 16.10:** Determine a sum formula for the cubes of numbers using Fermat's method (and for the sum of fourth powers of natural numbers).

## 16.6   Pascal's Method for Determining Formulas for the Sum of Powers

The French mathematician and philosopher Blaise Pascal (1623–1662) published the essay *Traité du triangle arithmétique* in 1654, in which he described the various properties of the particular triangle of numbers which is now called Pascal's triangle in his honor.

Starting from the binomial theorem for the third power, for example, he represented the difference between two consecutive cube numbers in two ways

$$(k+1)^3 - k^3 = \left(k^3 + 3k^2 + 3k + 1\right) - k^3 = 3k^2 + 3k + 1$$

and thus created a table (see Fig. 16.4).

If one sums up all numbers in the columns of Fig. 16.4, then in the 2nd column all summands except the two summands $(n+1)^3$ and $1^3$ vanish, so that we get the following for the sum of the first $n$ square numbers:

| k | $(k+1)^3 - k^3$ | $3k^2 + 3k + 1$ |
|---|---|---|
| 1 | $2^3 - 1^3$ | $3 \cdot 1^2 + 3 \cdot 1 + 1$ |
| 2 | $3^3 - 2^3$ | $3 \cdot 2^2 + 3 \cdot 2 + 1$ |
| 3 | $3^3 - 2^3$ | $3 \cdot 3^2 + 3 \cdot 3 + 1$ |
| ... | ... | ... |
| $n-1$ | $n^3 - (n-1)^3$ | $3 \cdot (n-1)^2 + 3 \cdot (n-1) + 1$ |
| $n$ | $(n+1)^3 - n^3$ | $3 \cdot n^2 + 3 \cdot n + 1$ |
| Sum | $(n+1)^3 - 1^3$ | $3 \cdot (1^2 + 2^2 + 3^2 + \ldots + n^2) + 3 \cdot (1 + 2 + 3 + \ldots + n) + n \cdot 1$ |

**Fig. 16.4**  Pascal's table for calculating the sum of cube numbers

$$3 \cdot \left(1^2 + 2^2 + 3^2 + \ldots + n^2\right)$$

$$= (n+1)^3 - 1^3 - 3 \cdot (1 + 2 + 3 + \ldots + n) - n \cdot 1$$
$$= n^3 + 3n^2 + 3n + 1 - 1 - 3 \cdot \tfrac{1}{2} \cdot \left(n^2 + n\right) - n$$

$$= n^3 + 3 \cdot \tfrac{1}{2} \cdot n^2 + \tfrac{1}{2} \cdot n$$

So:

$$1^2 + 2^2 + 3^2 + \ldots + n^2 = \tfrac{1}{3} \cdot n^3 + \tfrac{1}{2} \cdot n^2 + \tfrac{1}{6} \cdot n$$

**Suggestions for reflection and for investigations**
**A 16.11:** Determine the sum formula for cube numbers (and for the sum of fourth powers of natural numbers) according to Pascal's method.

## 16.7    Representation of the Sum Formulas using Bernoulli Numbers

The Swiss mathematician Jacob Bernoulli (1655–1705) discovered that in the sum formulas not only the binomial theorem and thus the binomial coefficients play a decisive role, but also certain other coefficients occur in all these formulas.

These are called Bernoulli numbers $B_k$ (with $k = 0$, 1, 2, 3, …) as suggested by the French mathematician Abraham de Moivre.

The sum formulas can then be noted as follows:

- **Sum of the first $n$ natural numbers**

$$S_1(n) = 1^1 + 2^1 + 3^1 + \ldots + n^1$$

$$= \tfrac{1}{2} \cdot \left[ \begin{pmatrix} 2 \\ 0 \end{pmatrix} \cdot B_0 \cdot n^2 + \begin{pmatrix} 2 \\ 1 \end{pmatrix} \cdot B_1 \cdot n^1 \right]$$

$$= \tfrac{1}{2} \cdot \left[ B_0 \cdot n^2 + 2 \cdot B_1 \cdot n^1 \right] = \tfrac{1}{2} \cdot \left[ n^2 + n^1 \right]$$

So: $B_0 = 1$ and $B_1 = \tfrac{1}{2}$.

- **Sum of the first $n$ square numbers**

$$S_2(n) = 1^2 + 2^2 + 3^2 + \ldots + n^2 = \tfrac{1}{3} \cdot \left[ \begin{pmatrix} 3 \\ 0 \end{pmatrix} \cdot B_0 \cdot n^3 + \begin{pmatrix} 3 \\ 1 \end{pmatrix} \cdot B_1 \cdot n^2 + \begin{pmatrix} 3 \\ 2 \end{pmatrix} \cdot B_2 \cdot n^1 \right]$$

$$= \tfrac{1}{3} \cdot \left[ B_0 \cdot n^3 + 3 \cdot B_1 \cdot n^2 + 3 \cdot B_2 \cdot n^1 \right] = \tfrac{1}{3} \cdot \left[ n^3 + \tfrac{3}{2} n^2 + \tfrac{1}{2} n^1 \right]$$

So: $B_0 = 1$, $B_1 = \tfrac{1}{2}$ and $B_2 = \tfrac{1}{6}$.

- **Sum of the first $n$ cube numbers**

$$S_3(n) = 1^3 + 2^3 + 3^3 + \ldots + n^3 = \tfrac{1}{4} \cdot \left[ \begin{pmatrix} 4 \\ 0 \end{pmatrix} \cdot B_0 \cdot n^4 + \begin{pmatrix} 4 \\ 1 \end{pmatrix} \cdot B_1 \cdot n^3 + \begin{pmatrix} 4 \\ 2 \end{pmatrix} \cdot B_2 \cdot n^2 + \begin{pmatrix} 4 \\ 3 \end{pmatrix} \cdot B_3 \cdot n^1 \right]$$

$$= \tfrac{1}{4} \cdot \left[ B_0 \cdot n^4 + 4 \cdot B_1 \cdot n^3 + 6 \cdot B_2 \cdot n^2 + 4 \cdot B_3 \cdot n^1 \right] = \tfrac{1}{4} \cdot \left[ n^4 + 2n^3 + n^2 \right]$$

So: $B_0 = 1$, $B_1 = \tfrac{1}{2}$, $B_2 = \tfrac{1}{6}$ and $B_3 = 0$.

---

**Suggestions for reflection and for investigations**

**A 16.12:** Show that Bernoulli's discovery also applies to the sum formulas for fourth and fifth powers, and thus determine the Bernoulli numbers $B_4$ and $B_5$.

## 16.8   Determination of Sum Formulas using Lagrange Polynomials

The French mathematician of Italian descent Joseph-Louis Lagrange (1736–1813)

created a formula with the help of which the term of a polynomial function can be derived if you know a certain number of points of the graph of that function.

This process of finding the term of a polynomial, whose graph passes through given points, is also called **interpolation.**

---

**Formula**

**Determination of a polynomial function whose graph passes through given points (Lagrange polynomial)**

The graph of a polynomial function of 2nd degree $L_2$ passes through three given points $P_0(x_0, y_0)$, $P_1(x_1, y_1)$, $P_2(x_2, y_2)$, whose function term can be calculated with the help of:

$$L_2(x) = \frac{(x - x_1)(x - x_2)}{(x_0 - x_1)(x_0 - x_2)} \cdot y_0 + \frac{(x - x_0)(x - x_2)}{(x_1 - x_0)(x_1 - x_2)} \cdot y_1 + \frac{(x - x_0)(x - x_1)}{(x_2 - x_0)(x_2 - x_1)} \cdot y_2$$

The graph of a polynomial function of 3rd degree $L_3$ passes through four given points $P_0(x_0, y_0)$, $P_1(x_1, y_1)$, $P_2(x_2, y_2)$, $P_3(x_3, y_3)$, whose function term can be calculated using:

$$L_3(x) = \frac{(x - x_1)(x - x_2)(x - x_3)}{(x_0 - x_1)(x_0 - x_2)(x_0 - x_3)} \cdot y_0 + \frac{(x - x_0)(x - x_2)(x - x_3)}{(x_1 - x_0)(x_1 - x_2)(x_1 - x_3)} \cdot y_1$$
$$+ \frac{(x - x_0)(x - x_1)(x - x_3)}{(x_2 - x_0)(x_2 - x_1)(x_2 - x_3)} \cdot y_2 + \frac{(x - x_0)(x - x_1)(x - x_2)}{(x_3 - x_0)(x_3 - x_1)(x_3 - x_2)} \cdot y_3$$

and so on.

*Note:* The approach assumes that all $x$-coordinates of the given points are different from each other. ◀

**Example: Determining a term for the sum of the first n natural numbers using polynomial interpolation**

The first three elements of the sum sequence of natural numbers are:

$$s(0) = 0, \ s(1) = 1, \ s(2) = 1 + 2 = 3$$

That means you know the points $P_0(0,0)$, $P_1(1,1)$, $P_3(2,3)$ through which a graph of 2nd degree, that is, a parabola, is supposed to pass (see following figure).

$$L_2(n) = \frac{(n-1)(n-2)}{(0-1)(0-2)} \cdot 0 + \frac{(n-0)(n-2)}{(1-0)(1-2)} \cdot 1 + \frac{(n-0)(n-1)}{(2-0)(2-1)} \cdot 3$$

$$= (-1) \cdot n \cdot (n-2) + \tfrac{3}{2} \cdot n \cdot (n-1) = \tfrac{1}{2} \cdot n^2 + \tfrac{1}{2} \cdot n$$

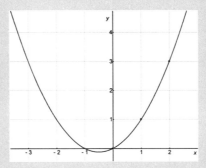

**Suggestions for reflection and for investigations**

**A 16.13:** Determine the sum formula for the sum of the first $n$ square numbers (cube numbers) using Lagrange's method of interpolation.

## 16.9   References to Further Literature

On **Wikipedia** you can find further information and literature on the keywords in English (German, French):

- Arithmetic progression (Arithmetische Folge/Arithmetische Reihe, Suite arithmé-tique/Somme arithmétique)
- Faulhaber's formula (Faulhaber'sche Formel, Formule de Faulhaber)
- Square pyramidal number (Quadratische Pyramidalzahl, Nombre pyramidal carré)
- Tetrahedral number (Tetraederzahl, Nombre tétraédrique)
- Polynomial interpolation (Polynominterpolation, Interpolation polynomiale)

Extensive informations can be found at **Wolfram Mathworld** under the keywords:

- Arithmetic Progression, Arithmetic Series, Faulhaber's Formula, Power Sum, Figurate Numbers, Pyramidal Number, Square Pyramidal Number, Lagrange interpolating Polynomial

A comprehensive historical account of the various approaches in deriving the formulae developed over the centuries is given in the article *Sums of Powers of Positive Integers* by Janet Beery (University of Redlands), published in the course of the 100th anniversary of the Mathematical Association of America (MAA 100):

- https://www.maa.org/press/periodicals/convergence/sums-of-powers-of-positive-integers.

# The Pythagorean Theorem

# 17

*The book of nature is written in the language of mathematics, and its characters are triangles, circles and other geometric figures.*

*(Galileo Galilei, Italian physicist and mathematician, 1564–1642)*

Perhaps the most famous theorem of geometry bears the name of the Greek mathematician Pythagoras of Samos, who lived in the sixth century B.C.

## 17.1 The Pythagorean Theorem and the Classical Proofs of Euclid

If you ask adults what memories they have of the *Pythagorean theorem* then you often only get the equation $a^2 + b^2 = c^2$ as an answer. Unfortunately, it has been forgotten (or never come to mind) that the theorem consists of a double statement. The converse of the proposition is at least as important in its application as the proposition itself. From the knowledge of the side lengths of a triangle one can deduce what kind of triangle is given (acute-angled, right-angled, or obtuse-angled).

In the following we will – as usual – call the vertices of a triangle with $A$, $B$, $C$ (counterclockwise), mark the adjacent inner angles with $\alpha$, $\beta$, $\gamma$ and the respective opposite sides of the triangle with $a$, $b$, $c$.

---

**Theorem**

**Pythagorean theorem**

If $\gamma$ is the right angle of a right-angled triangle with sides $a$, $b$, and hypotenuse $c$, then the relationship $a^2 + b^2 = c^2$ between the lengths of sides is valid.

However, we also have the **converse of the theorem:**

If the equation $a^2 + b^2 = c^2$ is satisfied by the side lengths of $a$, $b$, $c$ of a triangle, then the angle $\gamma$ which is opposite to the side $c$ is a right angle. ◀

---

A triple of integers $(a, b, c)$, which meet the condition $a^2 + b^2 = c^2$ is called a **Pythagorean triple,** see Sect. 2.7. An example is $(3, 4, 5)$, which is illustrated in the figure above.

It is not known whether Pythagoras or his disciples, the Pythagoreans, knew a reason or even a proof for the validity of the theorem. The first two documented proofs can be found in the *Elements* of Euclid (Book I, propositions 47 and 48, and Book VI, proposition 31).

### 17.1.1 First Proof by Euclid

Euclid's first proof of the theorem uses the property that two triangles are equal in area if they match in one side and in a corresponding altitude (*Elements* I, 38):

- *Triangles which are on equal bases and in the same parallels are equal to one another.*

To prove this, first divide the whole figure by a perpendicular from the point $C$ to side $c$ and extend this line.

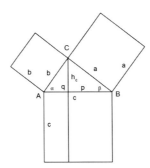

Then bisect one of the squares on a side by a diagonal and color one of the two isosceles right-angled triangles that have been generated. According to *Elements I, 38*, this bisected square has the same area as the obtuse triangle, which is colored in the second picture.

By rotating this triangle by 90° (clockwise), a congruent triangle is obtained in the third figure and finally, in the last figure, a right-angled triangle is to be seen with the same area, which bisects the rectangle that lies below the corresponding segment of the hypotenuse.

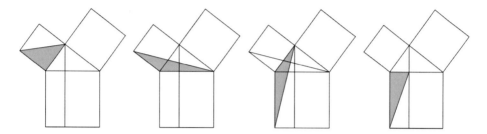

This proves that half of the square on a side and half of the corresponding rectangle below the hypotenuse have the same area, that is, the square on the side itself and the corresponding rectangle on the segment of the hypotenuse are of equal size.

Analogously, the proof for the square on the other side of the triangle can be performed.

If one designates the two shorter sides of the two rectangles, i.e. the segments of the hypotenuse, as usual with $q$ (vertical projection from $b$ upon $c$) and with $p$ (vertical projection from $a$ upon $c$), then this can be formulated as **Euclid's theorem**:

---

**Theorem**

**Euclid's theorem**

The square on a side of a right-angled triangle is equal in area to the rectangle which is determined by the corresponding segment of the hypotenuse and the hypotenuse itself:

$a^2 = c \cdot p$ and $b^2 = c \cdot q$.

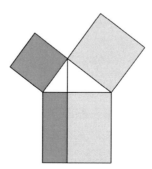

◄

This theorem is now applied to complete the proof of Pythagoras's theorem:

$$a^2 + b^2 = c \cdot p + c \cdot q = c \cdot (p + q) = c \cdot c = c^2$$

The proof of the converse of the Pythagorean theorem will be shown later (in Sect. 17.5).

## 17.1.2 Euclid's second proof

Book VI of the *Elements* deals with proportions in geometry. This is the basis of the second proof of the Pythagorean theorem. Theorem VI, 8 examines the above division of the right-angled triangle by the altitude $h_c$:

*If you draw the altitude $h_c$ in a right-angled triangle, then two right-angled subtriangles are created, which are similar to the original triangle:*

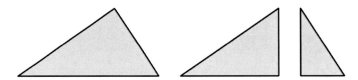

One can, therefore, set up a triple ratio equation for the side lengths of the three right-angled triangles (shorter side. longer side, hypotenuse). It holds:

**$a{:}b{:}c = h{:}q{:}b = p{:}h{:}a$**

The first proportion applies to the large (undivided) triangle, the second to the subtriangle, which is located to the left of the altitude $h_c$ and the third in the subtriangle, which is located to the right of the altitude $h_c$.

This triple ratio equation contains nine individual ratio equations, three of which can be converted into each other by changing the variables.

The equations can also be written down as products.

(1)   $a : b = h : q \Leftrightarrow a \cdot q = b \cdot h$   is consistent with

(9)   $q : b = h : a \Leftrightarrow a \cdot q = b \cdot h$

(2)   $a : b = p : h \Leftrightarrow a \cdot h = b \cdot p$   is consistent with

(6)   $h : b = p : a \Leftrightarrow a \cdot h = b \cdot p$

(3)   $h : q = p : h \Leftrightarrow h^2 = p \cdot q$

(4)   $a : c = h : b \Leftrightarrow a \cdot b = c \cdot h$   is consistent with

(8)   $b : c = h : a \Leftrightarrow a \cdot b = c \cdot h$

(5)   $a : c = p : a \Leftrightarrow a^2 = c \cdot p$

(7)   $b : c = q : b \Leftrightarrow b^2 = c \cdot q$

Statements (5) and (7) correspond to Euclid's theorem, from which – as explained above – the statement of the Pythagorean theorem follows. Statement (3) is the so-called **Euclid's right triangle altitude theorem** (see also theorem II, 14 of the *Elements*):

---

**Theorem**

**Euclid's right angle altitude theorem**

The square on the altitude $h_c$ of a right-angled triangle is equal in area to the rectangle formed by the hypotenuse segments $p$ and $q$.

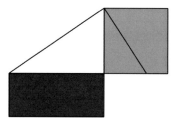

◄

*Annotation*: The theorem is also called **Geometric mean theorem** as the geometric mean of the two segments equals the altitude.

The statements (4) = (8) contain the two possibilities to calculate the area of a double triangle (that is, a rectangle): $a \cdot b = c \cdot h$.

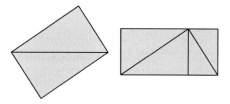

The statements (1) = (9) and (2) = (6) indicate relationships that are not important; these are illustrated in the following figures.

$$a \cdot q = b \cdot h \quad a \cdot h = b \cdot p$$

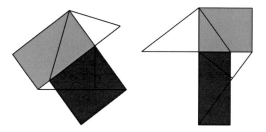

**Suggestions for reflection and for investigations**
**A 17.1:** Explain the following sequence of figures.

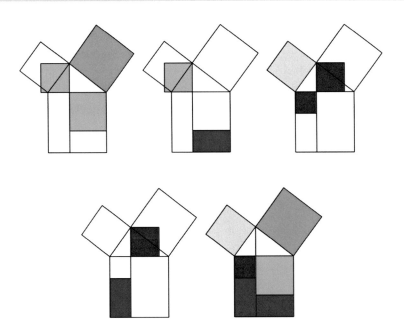

## 17.2   "Beautiful" Proofs of the Pythagorean Theorem

There are well over 100 proofs of the Pythagorean theorem – which one is probably the most beautiful?

Everyone must decide what is beautiful for herself or himself. One criterion could be: you just look and immediately see without any calculation, that together, the squares on the shorter sides have the same area as the square on the hypotenuse.

This criterion is certainly fulfilled by the evidence which Leonardo da Vinci (1452–1519) is said to have found: Complete the classical Pythagorean figure above and below each by the given right-angled triangle. At the top, an axially symmetrical hexagon is created; at the bottom a hexagon which is symmetrical to the center of the square on the hypotenuse. If you draw auxiliary lines you realize the subareas which are congruent.

The evidence presented in Fig. 17.1a, b. Isn't it beautiful?

**Fig. 17.1  a–c** Leonardo's proof at a glance (by courtesy of Dr. Peter Gallin, Switzerland) and the Italian stamp issued on the occasion of Leonardo's 500<sup>th</sup> birthday

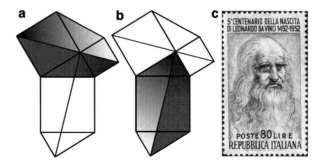

In the following figures, congruent subareas are of the same color so that it is easier to understand why the two hexagons have the same area.

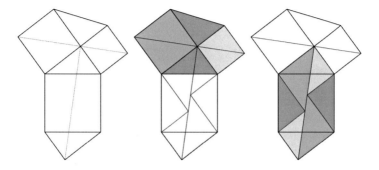

**Suggestions for reflection and for investigations**

**A 17.2:** In Fig. 17.2a the Pythagorean figure is completed by three colored triangles. Show that the following applies:

1. The three colored triangles each have the same area as the initial right-angled triangle.
2. If $x$, $y$, $z$, denote the longest sides of the colored triangles, then the following applies:

$$x^2 + y^2 + z^2 = 3 \cdot (a^2 + b^2 + c^2)$$

3. In Fig. 17.2b the figure from Fig. 17.2a is completed by squares. Show: The light blue-colored areas together are three times as large as the green-colored areas.

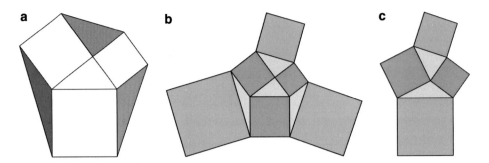

**Fig. 17.2  a–c** Completed and generalized Pythagorean figures

4. The properties (1) and (2) even apply to any triangle on whose sides squares are constructed.

**A 17.3:** Fig. 17.2c illustrates the following general theorem:
*If squares are drawn on the sides of an arbitrary triangle and if the area between two of the squares are completed to a triangle, on which a square is placed, then the two triangles (yellow) are equal in size and the total area of the two outer squares (light blue) is twice as large as the total area of the two inner squares (green).*

Examine which properties apply when the figures in Fig. 17.2b or Fig. 17.2c are complemented by further squares (to the right and to the left)?

## 17.3   Proofs of the Pythagorean Theorem by Dissection

Proofs by dissection are also *beautiful.* Here the squares on the sides of the right-angled triangle are dissected into subareas in such a way that the pieces can be used to fill both the squares on the two sides *and* the square on the hypotenuse.

### 17.3.1 Perigal's Proof by Dissection

In his spare time – he was a bank employee – the English amateur mathematician Henry Perigal (1801–1898), who was descended from French Huguenots, found a method for an infinite number of possible decompositions that need only five pieces. (By the way, he was so enthusiastic about his discovery that he had the symmetrical version of the dissection he had found chiseled onto his tombstone).

With this proof by dissection, the smaller of the two squares on the sides remains undivided, that is, it can be moved completely to the square on the hypotenuse. There are an infinite number of ways for the decomposition of the square on the other side.

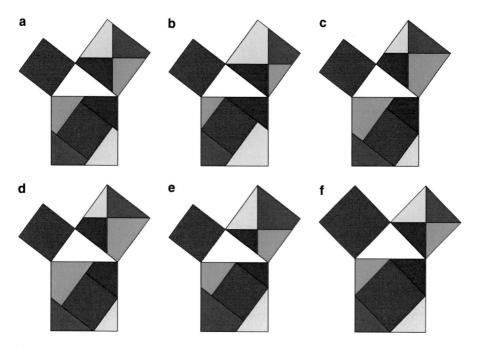

**Fig. 17.3 a–f** Various cases of Perigal's proof by dissection

The decomposition is done by a horizontal and a vertical line, which must run between opposite vertices of the square. To move the pieces from the square on the larger side to the square on the hypotenuse (and vice versa) only one shift is necessary (i.e., it is not necessary to turn or flip these pieces).

The horizontal and vertical lines inside the square on the greater side are each the same length as the hypotenuse.

In Fig. 17.3a the symmetrical position is chosen, which leads to the fact that all four pieces are congruent to each other. The next four decompositions in Fig. 17.3b–e represent extreme positions of the horizontal and vertical line respectively. In these extreme layers one could use two of the four pieces to cover the initial right-angled triangle.

Finally, Fig. 17.3f shows the special case of an isosceles right-angled triangle – both squares on the shorter sides are of equal size. Here we have only *one* method of decomposition: one of the two squares on the sides is divided by the diagonals into four congruent right-angled triangles.

## 17.3.2 Göpel's Proof by Dissection

This proof discovered by the German mathematician Adolph Göpel (1812–1847) also manages with a decomposition into five pieces which can be moved (without turning the pieces) from the squares on the sides to the square on the hypotenuse (and vice versa), see the following figures.

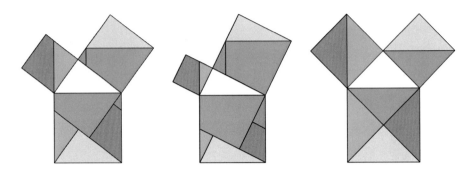

The square on the smaller side is divided into two parts by extending one vertical side of the square on the hypotenuse. The square on the larger side is divided into three pieces by a horizontal and a vertical line. (You could divide the larger square by extending the other vertical side of the hypotenuse square, but then you would have to rotate pieces when moving them).

The yellow-colored piece is the same size as the given right-angled triangle, the gray-blue colored piece and the light-blue colored piece are each similar to the initial triangle, that is, their side lengths result from simple ratio equations.

The figure on the left shows the decomposition for $a : b = 4 : 3$, the central figure for $a = 2 \cdot b$ (with congruent gray-blue and light-blue colored pieces) and the figure on the right shows the special case for $a = b$.

### 17.3.3 Gutheil's Proof by Dissection

For the decomposition of the squares on the sides and the hypotenuse according to Benjir von Gutheil († 1914), one generally needs seven pieces.

What is fascinating about this decomposition is the symmetrical arrangement of the pieces: in the square on the hypotenuse, the pieces are arranged symmetrically with respect to the center of the square, in the squares on the sides they are axially symmetrical. The square on the smaller side is divided by a diagonal into two isosceles right-angled triangles. The square on the larger side is made up of two pieces that match the given right-angled triangle, a squarish piece with side length $a - b$ and two triangular pieces that complement the two triangles of the decomposition of the square on the smaller side to form the initial triangle.

Since the square on the hypotenuse is decomposed in a centrally symmetrical way and the square on the side in an axially symmetrical way, you have to flip two of the pieces when moving from the square on a shorter side to the square on the hypotenuse.

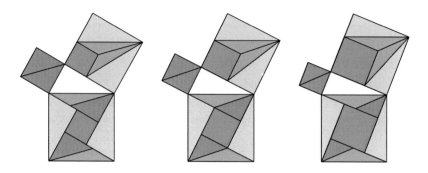

### 17.3.4 Epstein's and Nielsen's Proof by Dissection

Eight pieces are needed for the decomposition discovered by Paul Epstein (1871–1939) and Jakob Nielsen (1890–1959). These are obtained by dividing the squares on the sides by the diagonals and by drawing lines parallel to the hypotenuse of the initial triangle through the other two vertices of the squares.

Since both sides are divided in the same way, corresponding pieces in both sides are similar to each other. The square on the hypotenuse is divided by the bisecting line of the right angle; the pieces are arranged in such a way that a point-symmetrical decomposition of the square on the hypotenuse is generated. When drawing the pieces, note that the angles in the acute-angled triangles are $\alpha$, $45°$ and $135° - \alpha$.

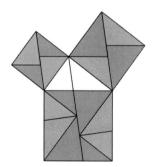

**Suggestions for reflection and for investigations**
**A 17.4:** What is the peculiarity of the pieces in the proof of Epstein and Nielsen if the initial triangle is isosceles right-angled?

### 17.3.5  Dobriner's and Thieme's Proof by Dissection

The decomposition by Hermann Dobriner (1857–1902) and Karl Gustav Hermann Thieme (1852–1926) is carried out in such a way that Euclid's theorem is also proved. That is, the pieces by which the square on the side $a$ is dissected fit exactly into the rectangle with the sides $p$ and $c$ and those by which the square on the side $b$ is dissected fit exactly into the rectangle with the sides $q$ and $c$ and vice versa. The basic idea of the decomposition is to divide the squares on the sides into strips of the width $p$ and $q$ and dissect the resulting pieces by horizontal lines so that the rectangular areas of the square on the hypotenuse can be filled.

The number of pieces for the square on the larger non-hypotenuse side is always three: The right vertical side of the square on the hypotenuse is extended so that a piece is cut off that has the same size as the initial triangle. The remaining square is divided by a line through vertex $C$ that is parallel to the hypotenuse so that the remaining piece (light blue) fits into the right rectangle below the hypotenuse.

In the case of an isosceles right-angled triangle, only three pieces are needed for the square on the smaller side – but in general, the number of pieces depends on the width of the segment of the hypotenuse and in principle it can be arbitrarily large.

The first two figures show the decompositions for the case $a : b = 4 : 3$, the others for the case $a : b = 2 : 1$ (exactly three stripes).

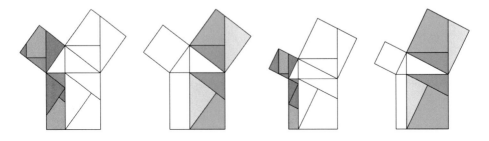

### 17.4    Presentation of Proofs by Means of Tile Patterns

Anyone who has ever thought about how to tile a kitchen or a bathroom knows that you can "tessellate" a plane with squares. Such a pattern (we will call it a S-pattern, see figure on the left) is, however, quite boring.

More diversified is a pattern of two squares of different sizes, that is, of small and large squares (we will call it a s&l-pattern, see the following figure on the right), with which a plane can also be completely filled.

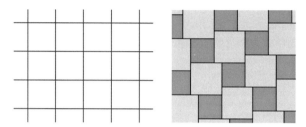

In every *s&l*-pattern a *S*-pattern is hidden: because if you select an arbitrary point in the large or the small squares of the *s&l*-pattern and connect these points with each other, then you get exactly a *S*-pattern.

This can be implemented as follows: an area is tiled by squares of different sizes (*s&l*-pattern). Then you put a transparent film over it, on which you mark one special point in each *s&l*-element and connect these points.

Depending on your choice of points, you get the decomposition according to Perigal or according to Göpel. The yellow and green tiles form the squares on the sides of a right-angled triangle, and the squares of the S-pattern on the film form the squares on the hypotenuse.

The figure on the left shows the symmetrical case of the Perigal proof: the smaller square of the side (green) is not divided by the red lines, the larger one (yellow) is cut into pieces by the red lines – the square on the hypotenuse (formed by the red lines) is cut into one green and four yellow pieces. The film can be moved left or right or up or down as long as only the green square remains within the red frames – compare the explanation of Perigal's idea.

In the figure on the right, the red lines divide the squares on the sides into two or three pieces – the square on the hypotenuse (formed by the red lines) is divided into three yellow and two green puzzle pieces. The film must be placed in such a way that the vertices of the red framed square on the hypotenuse correspond to the lower points, where that belong to adjacent yellow and green colored squares.

That this is indeed a proof of the Pythagorean theorem, that is *yellow + green = framed red*, results from the fact that with both patterns the plane is completely tessellated.

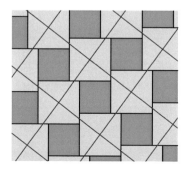
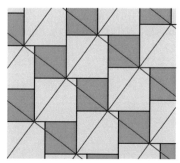

## 17.5   Some Proofs of Historical Significance

As mentioned above, there are well over 100 proofs of the Pythagorean theorem.

- The Greek mathematician Proclus (412–485 A.D.) was convinced that the idea for
  the following two figures came from Pythagoras himself, and called this **proof of
  Pythagoras.** That this is indeed a proof of the theorem, becomes clear when one considers what is represented by the two figures and what is the difference between the figures.

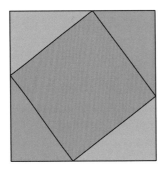 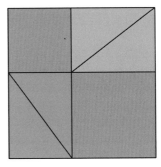

This is not so easy to see in the next figure.

- In a manuscript of **Chou Pei Suan Ching,** which was written in China before
  the first century B.C., the following figure can be found. The square with the side
  length $c = 5$ is obtained when you cut off four right-angled triangles from the outer
  square which has the side length $a + b = 3 + 4$. These triangles have an area of
  $\frac{1}{2} \cdot a \cdot b = \frac{1}{2} \cdot 3 \cdot 4 = 6$ units each. Applying the binomial theorem for $(a + b)^2$ you
  get the following relationship between the square with side length $c$ and the two (not
  drawn) squares with the side lengths $a$ and $b$:

$$c^2 = (a + b)^2 - 4 \cdot \frac{1}{2} \cdot a \cdot b = a^2 + b^2$$

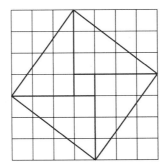

- The following proof originates from the Indian mathematician **Bhaskara** (1114–1185): a square with the side length $c$ is dissected into four right-angled triangles with the sides $a$ and $b$ and a square with the side length $b - a$ ($a < b$). Then it follows:

$$c^2 = 4 \cdot \frac{1}{2} \cdot a \cdot b + (b - a)^2 = 2ab + b^2 - 2ab + a^2 = a^2 + b^2$$

Bhaskara writes that one only has to look at this proof ("See!"), but without knowledge of algebra (binomial theorem) one will not get far here either.

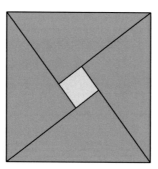

- Three **Proofs of Thabit ibn Qurra** (836–901): The eminent Assyrian mathematician and astronomer was originally active as a money changer before he was invited by the Banu Musa brothers, due to his gift for languages, to translate the writings of Greek mathematicians in the "House of Wisdom" in Baghdad. After having studied Euclid's *Elements* intensively he independently developed proofs for the Pythagorean theorem by himself.

In his first proof, he drew the two squares on the sides $a$ and $b$ ($a > b$) side by side. Into this figure the right-angled triangle with the sides $a$ and $b$, and the hypotenuse $c$ can be entered twice (light blue and green). Then at the third step, both triangles are turned upward (counterclockwise or clockwise) around the upper left and right vertices of the squares, to form the square on the hypotenuse.

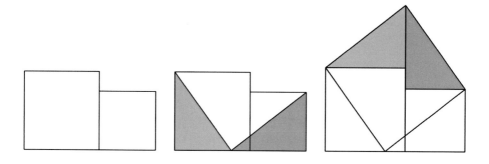

The ingenuity of the approach to the second proof of Thabit ibn Qurra is particularly evident if one colors the figure differently or if one turns a part of the figure by 180° (see reduced figure on the right).

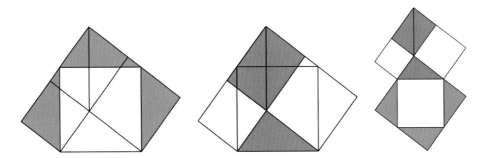

The third proof deals with the generalization of the Pythagorean theorem and thus also contains a **proof of the converse of the theorem.**

In an obtuse triangle *ABC* on the right the vertices *D* and *E* are plotted on the segment *AB* so that the triangles *ABC, ACD,* and *CBE* are similar to each other.

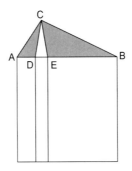

Therefore, we get the following:
$|AC| : |AB| = |AD| : |AC|$, so $|AC|^2 = |AB| \cdot |AD|$, and
$|BC| : |AB| = |EB| : |BC|$, so $|BC|^2 = |AB| \cdot |EB|$.
Thus we get: $|AC|^2 + |BC|^2 = |AB| \cdot (|AD| + |EB|)$.
Is the angle at *C* a right angle, then the vertices *D* and *E* coincide, and it follows that:

$$|AC|^2 + |BC|^2 = |AB|^2$$

If the angle at *C* is obtuse, then the square on the side *AB* must be reduced by the rectangle of width *DE* so that it is equal in area to the sum of the squares on the shorter sides:
$|AC|^2 + |BC|^2 = |AB|^2 - |AB| \cdot |DE|$
In the case of an acute-angled triangle, the square on the side *AB* must be enlarged.

**Suggestions for reflection and for investigations**
**A 17.5:** Explain the following proof described by Abu'l-Wafa Al-Buzjani (940–998), one of the most important Persian mathematicians of the Middle Ages: Two squares with side lengths $a$ (yellow) and $b$ (white), where $a < b$, are dissected into pieces according to the figure in the center and then put together to form the figure on the right.

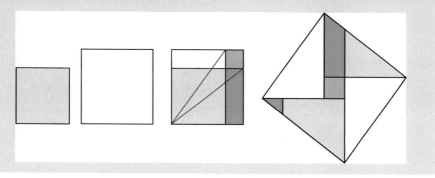

## 17.6   Infinite Pythagorean Sequences

There is still much to discover about the Pythagorean theorem ….

To draw a Pythagorean **spiral** one begins with a right-angled triangle with sides of equal length of 1 unit. The hypotenuse thus has the length $\sqrt{2}$. This hypotenuse then becomes one of the two sides of the next right-angled triangle. The other side has again the length 1, so that the new hypotenuse has the length $\sqrt{3}$ and so on.

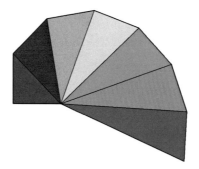

**Suggestions for reflection and for investigations**
**A 17.6:** Explore:

1. What is the area of the growing Pythagorean spiral after $n$ steps?
2. How many triangles can be drawn until the triangles of the Pythagorean spiral overlap?

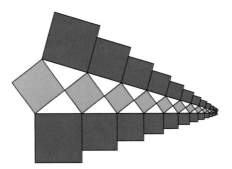

In the following Pythagorean sequence the classical Pythagorean figure is first reflected on the diagonal that passes through the two squares on the sides and then the figure is continued by the square on the smaller side becoming the square on the larger side.

**A 17.7:** Examine the area of the growing figure and how this depends on the initial sizes.

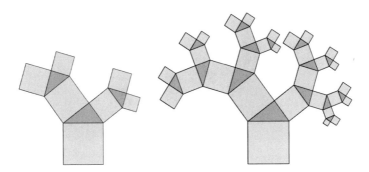

The **Pythagorean tree** starts with the classical Pythagorean figure, but then use each of the squares on the sides as squares on the hypotenuse in the following steps.

**A 17.8:** Examine the area of the growing figure, the side lengths of the sides and the step where parts of the figure overlap.

## 17.7 Generalization of the Pythagorean Theorem

The theorem about the equality of area of the squares on the sides and the hypotenuse of a right-angled triangle does not only apply to squares but also to any similar figures.

Because the calculation of the area $A$ of all figures are done using formulas of the type $A = k \cdot s^2$ where $s$ stands for the side lengths of $a$ or $b$ or $c$ and $k$ is a factor which is characteristic to the figure:

$k = 1$ applies for squares, $k = \frac{\sqrt{3}}{4}$ for equilateral triangles, $k = \frac{\sqrt{25+10\sqrt{5}}}{4}$ for regular pentagons, $k = \frac{3\sqrt{3}}{2}$ for regular hexagons, $k = 2 \cdot \left(\sqrt{2}+1\right)$ for regular octagons and $k = \frac{\pi}{8}$ for semicircles on the sides of the right-angled triangle.

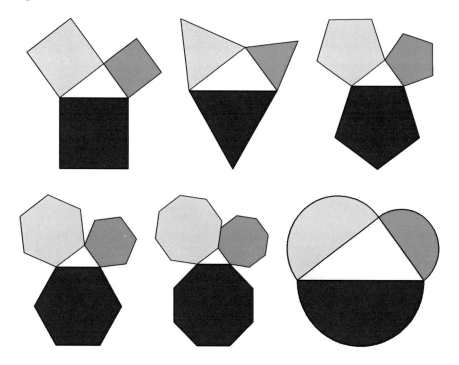

## 17.8 The Lune of Hippocrates of Chios and Other Circle Figures

The last figure in Sect. 17.7 leads to an astounding theorem discovered by the Greek mathematician Hippocrates of Chios around 450 B.C.: straight-line limited areas can be exactly as large as curvilinearly limited areas. In fact:

**Theorem**

## The Lune of Hippocrates

The area of a right-angled triangle is equal to the sum of the areas of the two lunes above the sides.

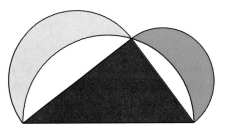

◄

To prove the theorem, one only has to apply the idea of the generalized Pythagorean theorem from Sect. 17.7: The area of the semicircle on the hypotenuse of a right-angled triangle is equal to the sum of the areas of the semicircles on the sides (see the following figure on the left). If you then fold up the semicircle "below" the hypotenuse (see the following figure on the right), you will see the equality.

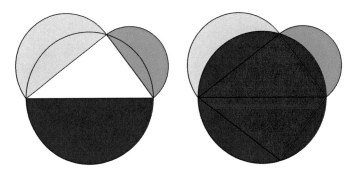

**Suggestions for reflection and for investigations**

**A 17.10:** Explain the two graphics below. In each case, describe the facts in words.

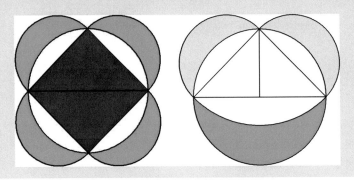

**A 17.11:** Prove a theorem about the lunes over regular hexagons, also found by Hippocrates.

*The area of the six lunes together with the area of the small circle is the same as the area of the yellow-colored area within the regular hexagon. The green area within the hexagon is a circle whose radius is half the length of the sides of the hexagon.*

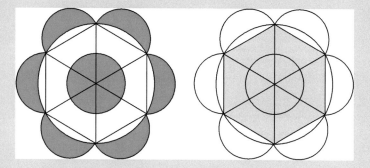

**A 17.12:** The idea of the proof for the lunes of Hippocrates can also be applied to the following **heart figure** – not only for circular arcs but also for regular $n$-sided polygons, where $n$ must be a multiple of four. Explain this.

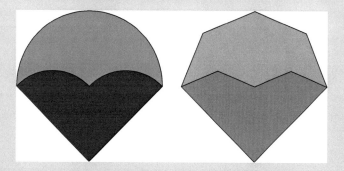

**A 17.13:** What is the relationship between the areas of the semicircles to be seen in the following figure?

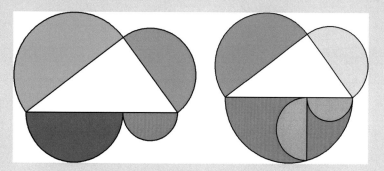

**A 17.14:** Prove that the (yellow colored) sickle-shaped figure, the so-called **Archimedes' shoemaker's knife** (Greek: *Arbelos*), is equal in area to the red colored circle. The diameter of the circle is the same length as the marked altitude.

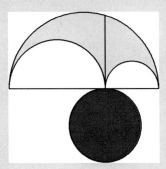

*Hint:* Use the following auxiliary figures for the demonstration.

**A 17.15:** Into the two partial areas of the shoemaker's knife from A 17.14 two circles can be sketched in (so-called **twin circles of Archimedes**). Why are these two green-colored circles of the same size?

*Hint:* The Italian stamp was issued on the occasion of the Archimedean year in 2013 and shows, among other things, the twin circles.

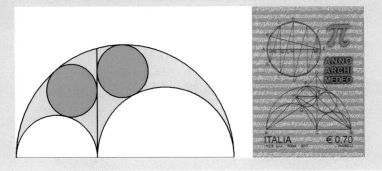

**A 17.16:** In his *Book of Lemmas* Archimedes describes a figure formed by four semicircles. Because of its shape it is called **Salinon** (in English: *salt cellar*). Explain why the green colored area in the figure on the left is of the same size as the area of the light blue colored circle in the central figure. How does this result in the equality of the differently colored areas in the figure on the right?

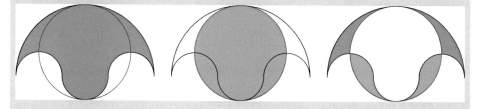

**A 17.17:** Among the numerous variations of the lunes that Hans Walser has investigated, there is also a sequence of lunes that are formed over nested isosceles right-angled triangles. Each lune is half the size of its predecessor.

Explain the construction of the sequence of lunes.

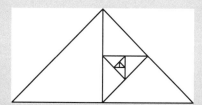

Explain the following theorem:

If you draw an initial triangle with an area of 1 unit, then the infinite number of lunes together have the area $1 + \frac{1}{2} + \frac{1}{4} + \frac{1}{8} + \frac{1}{16} + \frac{1}{32} + \ldots = 2$.

## 17.9   Application of the Pythagorean Theorem to Quadrilaterals

You can also draw squares on the sides of a quadrilateral. A remarkable fact applies here:

---

**Theorem**

**Orthodiagonal quadrilaterals**
The sums of the squares on opposite sides of a quadrilateral are equal if and only if the diagonals of the quadrilateral intersect orthogonally. ◄

---

As is well-known, a quadrilateral is not uniquely determined by the lengths of the four sides $a$, $b$, $c$, $d$; for four given side lengths, infinitely many quadrilaterals exist. But the theorem says:

No matter what shape the quadrilaterals have: if the sides meet the condition $a^2 + c^2 = b^2 + d^2$, then the diagonals intersect at right angles (and vice versa).

*For the proof:* If the diagonals in a quadrilateral intersect orthogonally, then we have (see Fig. 17.4b).
$$a^2 = u^2 + y^2 \ and \ c^2 = x^2 + v^2 \quad and \quad b^2 = y^2 + v^2 \quad and \quad d^2 = x^2 + u^2 \quad that \quad is:$$
$a^2 + c^2 = b^2 + d^2$.

If on the other hand the diagonals of a quadrilateral intersect at the angles $\varepsilon$ and $180° - \varepsilon$, then the following applies according to the Law of cosines:
$$a^2 = u^2 + y^2 - 2uy \cdot \cos(\varepsilon) \text{ and } c^2 = v^2 + x^2 - 2vx \cdot \cos(\varepsilon) \text{ that is}$$
$$a^2 + c^2 = u^2 + v^2 + x^2 + y^2 - 2 \cdot (uy + vx) \cdot \cos(\varepsilon) \text{ and in addition}$$
$$b^2 = v^2 + y^2 - 2vy \cdot \cos(180° - \varepsilon) \text{ and } d^2 = x^2 + u^2 - 2xu \cdot \cos(180° - \varepsilon) \text{ that is:}$$

$$b^2 + d^2 = u^2 + v^2 + x^2 + y^2 + 2 \cdot (vy + ux) \cdot \cos(\varepsilon)$$

From $a^2 + c^2 = b^2 + d^2$, therefore, follows $(uy + vx + vy + ux) \cdot \cos(\varepsilon) = 0$ and, since not all the summands $uy$, $vx$, $vy$, $ux$ can be zero at the same time, this results in $\cos(\varepsilon) = 0$, that is $\varepsilon = 90°$.

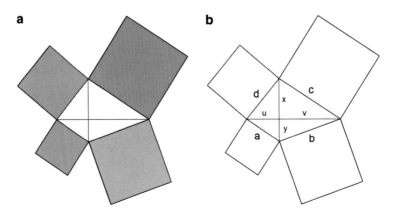

**Fig. 17.4   a, b** Orthodiagonal quadrilaterals

The following experiment is particularly impressive: Make strips of solid material (cardboard, wood, metal, …) of the lengths $a$, $b$, $c$, $d$, which meet the condition $a^2 + c^2 = b^2 + d^2$. These are connected to each other at the ends (e.g., with a clip) so that they can revolve and they form a quadrilateral. The diagonals can then be realized by rubber bands. Whatever shape the quadrilateral has: The rubber bands cut at right angles!

**Suggestions for reflection and for investigations**
**A 17.18:**

1. A special case of the theorem is given if $y = 0$ (see the following figure). What special characteristic results?
2. Another special case is given if, for example, $x = y$. What kind of figure is then present?

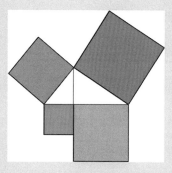

## 17.10  Integral Pythagorean partners and special Pythagorean sequences

The right-angled triangle with sides of integral lengths 1 and 7 has a hypotenuse which is the same length as the hypotenuse of the isosceles right-angled triangle with two sides of the integral length 5:

$50 = 1^2 + 7^2 = 5^2 + 5^2$ (see Fig. 17.5a)

There are also two different integer partners for the square on the hypotenuse when the area is 65 and 85:

$65 = 1^2 + 8^2 = 4^2 + 7^2$ (see Fig. 17.5b) and
$85 = 2^2 + 9^2 = 6^2 + 7^2$ (see Fig. 17.5c)

**Suggestions for reflection and for investigations**
**A 17.19:** Determine another ten such integral Pythagorean partners.

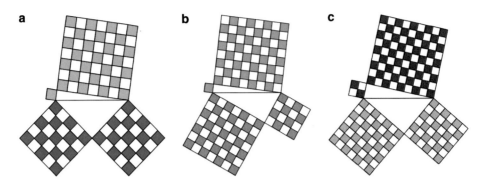

**Fig. 17.5  a–c** Integral Pythagorean partners to the sums 50, 65, 85

In Sect. 2.7 a descriptive proof for the Pythagorean triples (3, 4, 5) was given by means of colored stones. Another idea is presented in the following figure: The square with side length 3 is dissected in such a way that the parts of it together with the undissected square of side length 4 result in a decomposition of the square of side length 5.

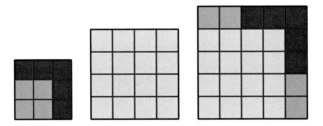

As well as the equation $3^2 + 4^2 = 5^2$ which describes a special relationship between the squares of the three consecutive natural numbers 3, 4, 5, there are other (even infinitely many) relations of a similar kind. The next equation of this type is

$$10^2 + 11^2 + 12^2 = 13^2 + 14^2$$

Here, too, a proof is possible by dissection: The two undissected squares with side lengths 11 and 12 can be supplemented by pieces from the square with side length 10 so that you get squares with side lengths 13 and 14 (see Fig. 17.6).

Also the next equations of this kind

$$21^2 + 22^2 + 23^2 + 24^2 = 25^2 + 26^2 + 27^2$$
$$36^2 + 37^2 + 38^2 + 39^2 + 40^2 = 41^2 + 42^2 + 43^2 + 44^2$$

can be proved by subdividing the smallest square and rearranging to larger squares.

Hans Walser proves the equations by using a grid of equilateral triangles (see Fig. 17.7). He uses the property that, according to the formula (2.2), an equilateral triangle with the side length $n$ can be composed of $n^2$ equilateral triangles of side length 1 (see also the following figure).

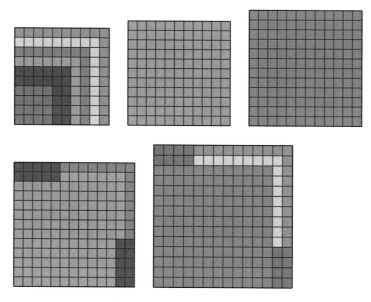

**Fig. 17.6**   Proof of the equation $10^2 + 11^2 + 12^2 = 13^2 + 14^2$ by dissection

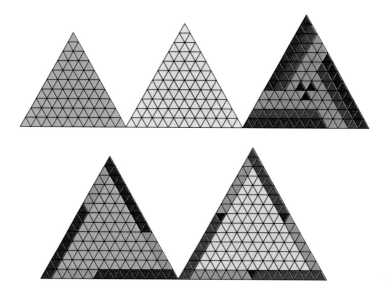

**Fig. 17.7**   Alternative presentation of the equation $10^2 + 11^2 + 12^2 = 13^2 + 14^2$ by means of equilateral triangles (with the friendly permission of Dr. Hans Walser, Switzerland)

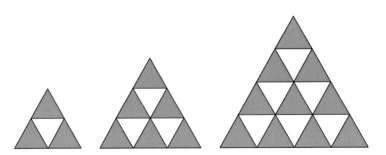

However, these relationships can also be represented graphically – similar to the method described above (see Fig. 17.8).

How can we now find out which other equations of this kind apply?

Possible observations:

- The two square numbers standing immediately to the left and right of the equal sign belong to special Pythagorean triples $(a, b, c)$ with $c = b + 1$:
  $(3, \mathbf{4}, \mathbf{5})$, $(5, \mathbf{12}, \mathbf{13})$, $(7, \mathbf{24}, \mathbf{25})$, $(9, \mathbf{40}, \mathbf{41})$. The next triple of this kind is $(11, \mathbf{60}, \mathbf{61})$.
- The number of summands on the left and right side of the equations increases by 1 each time, that is, in the next equation, there should be six summands on the left and five on the right.
- The bases of each *smallest* summands on the left side of the equations belongs to an arithmetic sequence of the $2^{nd}$ order: The sequence of differences of 3, 10, 21, 36 is the sequence 7, 11, 15. Accordingly, the left side of the next equation should start with the summand $55^2$.

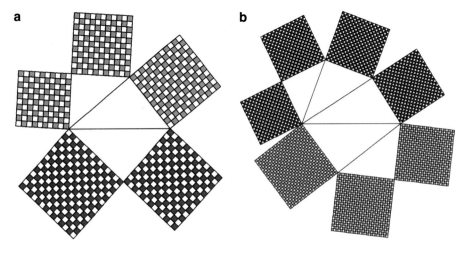

**Fig. 17.8  a, b** Illustration of the equations $10^2 + 11^2 + 12^2 = 13^2 + 14^2$ and $21^2 + 22^2 + 23^2 + 24^2 = 25^2 + 26^2 + 27^2$

- The bases of each *biggest* summands on the left side of the equations also belong to a arithmetic sequence of the 2$^{\text{nd}}$ order: The sequence of differences of 4, 12, 24, 40 is the sequence 8, 12, 16. Accordingly, the left side of the next equation should end with the summand 60$^2$.

*Intermediate result for the left side of the equation:*
The sum $55^2 + 56^2 + 57^2 + 58^2 + 59^2 + 60^2 = 19,855$ contains six summands.

- The bases of each *smallest* summands on the right side of the equations belong to an arithmetic sequence of the 2$^{\text{nd}}$ order: The sequence of the differences of 5, 13, 25, 41 is the sequence 8, 12, 16. Accordingly, the right side of the next equation should start with the summand 61$^2$.
- The bases of each *biggest* summands on the right side of the equations also belong to an arithmetic sequence of the 2$^{\text{nd}}$ order: The sequence of differences of 5, 14, 27, 44 is the sequence 9, 13, 17. Accordingly, the right side of the next equation should end with the summand 65$^2$.

*Intermediate result for the right side of the equation:*
The sum $61^2 + 62^2 + 63^2 + 64^2 + 65^2 = 19,855$ contains five summands.
So left and right side are the same.

**Generalization:** Suitable natural numbers $n$ are investigated, so that for $k = 1, 2, 3, \ldots$ we get:

$$(n - k)^2 + (n - k + 1)^2 + \ldots + n^2 = (n + 1)^2 + (n + 2)^2 + \ldots + (n + k)^2$$

This condition is satisfied if:

$$n = 2 \cdot k \cdot (k + 1) = 2k^2 + 2k$$

(See also the examples above: $n$ is the largest number on the left side of the equation:
$k = 1 \rightarrow n = 4$;  $k = 2 \rightarrow n = 12$;  $k = 3 \rightarrow n = 24$;  $k = 4 \rightarrow n = 40$;
$k = 5 \rightarrow n = 60$)
The proof is provided by the calculation:

$$(n - k)^2 + (n - k + 1)^2 + \ldots + n^2 = (n + 1)^2 + (n + 2)^2 + \ldots + (n + k)^2$$
$$\Leftrightarrow [1^2 + 2^2 + \ldots + n^2] - [1^2 + 2^2 + \ldots + (n - k - 1)^2]$$
$$= [1^2 + 2^2 + \ldots + (n + k)^2] - [1^2 + 2^2 + \ldots + n^2]$$
$$\Leftrightarrow 2 \cdot [1^2 + 2^2 + \ldots + n^2] = [1^2 + 2^2 + \ldots + (n - k - 1)^2]$$
$$+ [1^2 + 2^2 + \ldots + (n + k)^2]$$

Using the sum formula (2.4) and multiplying both sides of the equation by 6 results in

$$2 \cdot n \cdot (n + 1) \cdot (2n + 1)$$
$$= (n - k - 1) \cdot (n - k) \cdot (2n - 2k - 1) + (n + k) \cdot (n + k + 1) \cdot (2n + 2k + 1)$$

For $n = 2k^2 + 2k$ you get a true statement (general equation):

$$2 \cdot (2k^2 + 2k) \cdot (2k^2 + 2k + 1) \cdot (4k^2 + 4k + 1)$$
$$= (2k^2 + 2k - k - 1) \cdot (2k^2 + 2k - k) \cdot (4k^2 + 4k - 2k - 1)$$
$$+ (2k^2 + 2k + k) \cdot (2k^2 + 2k + k + 1) \cdot (4k^2 + 4k + 2k + 1)$$
$$\Leftrightarrow 2 \cdot (2k^2 + 2k) \cdot (2k^2 + 2k + 1) \cdot (4k^2 + 4k + 1)$$
$$= (2k^2 + k - 1) \cdot (2k^2 + k) \cdot (4k^2 + 2k - 1)$$
$$+ (2k^2 + 3k) \cdot (2k^2 + 3k + 1) \cdot (4k^2 + 6k + 1)$$
$$\Leftrightarrow 32k^6 + 96k^5 + 120k^4 + 80k^3 + 28k^2 + 4k$$
$$= 32k^6 + 96k^5 + 120k^4 + 80k^3 + 28k^2 + 4k$$

Alternatively, you can also solve the first equation for $n$ and then get the condition $n = 2k^2 + 2k$.

## 17.11  Heronian Triangles

In geometry, triangles, in which all side lengths and the area are integers, are called **Heronian triangles** in honor of the Greek mathematician Heron of Alexandria (around 10–70 A.D.).

Right-angled triangles whose side lengths form a Pythagorean triple are also Heronian triangles. This is because at least one of the sides has an *even* side length; therefore the area of the triangle which is half the product of the two sides is also an integer.

Heronian triangles can also be obtained by putting two right-angled triangles with integral side lengths together which match in one side – hence they are called **decomposable Heronian triangles**. The composition can be done in two "directions".

**Example 1**
The right-angles triangles belonging to the triples (5, **12**, 13) and (9, **12**, 15) each have a side with the side length of **12**.

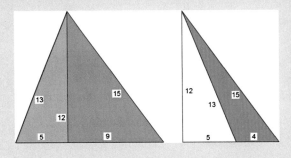

**Example 2**

The right-angled triangles belonging to the triples (5, **12**, 13) and (**12**, 16, 20) each have a side with the side length **12**.

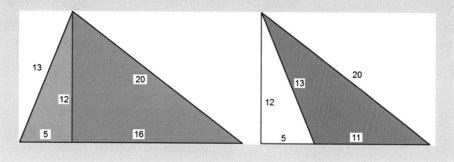

**Example 3**

The right-angled triangles belonging to the triples (**15**, 20, 25) and (**15, 36, 39**) each have a side with the side length **15**.

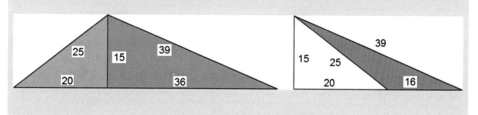

**Example 4**

The right-angled triangles belonging to the triples (15, **20**, 25) and (**20**, 48, 52) each have a side with the side length **20**.

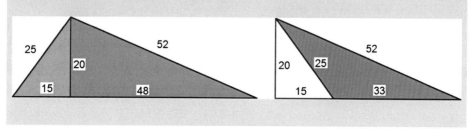

The area $A$ of the compound triangle is calculated in all examples from $A = \frac{1}{2} \cdot z \cdot h$ where $h$ is the length of the common side and $z$ is the sum or the difference of the lengths of the non-common side.

In general eight Heronian triangles exist to each of the two Pythagorean triples:

For any primitive (i.e., mutually different) Pythagorean triples $(a, b, c)$ and $(d, e, f)$ there exist an infinite number of mutually similar right-angled triangles with integer multiples of the original side lengths, that is, the $r$-times the side lengths $a, b, c$ or the $s$-times the side lengths $d, e, f$.

The factors $r, s$ can be chosen – with the help of the least common multiple – in such a way that the two triangles correspond in the lengths of two sides each:

$$(1) \quad r \cdot a = s \cdot d \quad (2) \quad r \cdot a = s \cdot e \quad (3) \quad r \cdot b = s \cdot d \quad (4) \quad r \cdot b = s \cdot e$$

**Suggestions for reflection and for investigations**

**A 17.20**

1. Explain which primitive Pythagorean triples and which multiples are considered in examples 1 to 4.
2. Specify which Heronian squares result from the combination of the number triples $(3, 4, 5)$ with the triple $(8, 15, 17)$.
3. Specify which Heronian squares result from the combination of the triple $(3, 4, 5)$ with the triple $(7, 24, 25)$.

Apart from Heronian triangles, which can be easily determined, there are also triangles that cannot not be obtained by putting two right-angled triangles together (i.e., as sum or difference). The first triangle of this kind was found by the American mathematician Fitch Cheney in 1929; it has the side lengths 25, 34, and 39. The smallest triangle of this type has the side lengths 5, 29, 30.

The general theory of this requires a deeper knowledge of geometry, which cannot be discussed here. In 2001 the following theorem was finally proved:

**Theorem**

**Non-decomposable Heronian triangles**

The vertices of non-decomposable Heronian triangles can be drawn in a square grid. That is, the vertices of these triangles have integer coordinates.

Non-decomposable Heronian triangles thus appear as residual triangles when three right-angled triangles with integral side lengths are cut off from suitable rectangles.

The areas of the triangles shown below are:

$$A = 30 \cdot 36 - \tfrac{1}{2} \cdot (30 \cdot 16 + 15 \cdot 20 + 15 \cdot 36) = 420$$
$$A = 21 \cdot 24 - \tfrac{1}{2} \cdot (21 \cdot 20 + 3 \cdot 4 + 18 \cdot 24) = 72$$
$$A = 30 \cdot 28 - \tfrac{1}{2} \cdot (30 \cdot 16 + 9 \cdot 12 + 21 \cdot 28) = 252$$

◄

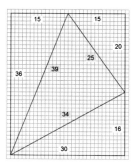
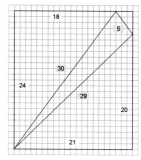
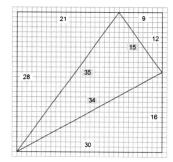

## 17.12   Stamps of Pythagoras and the Pythagorean Theorem

The postal administrations of some countries issued stamps dealing with Pythagoras and the theorem named after him:

Nicaragua (1971), South Korea (2014), Suriname (1972)

Greece (1955), Macedonia (1998)

Vatican (1986), Sierra Leone (1983)

Philippines (2001), Japan (1984), San Marino (1983)

## 17.13  References to further literature

On Wikipedia you can find further information and literature in English (German, French) about the keywords:

- Pythagorean theorem* (Satz des Pythagoras*, Théorème de Pythagore*)
- Orthodiagonal quadrilateral (Orthodiagonales Viereck, -)
- Pythagorean tiling (only in English)
  *) Marked as readable/excellent article

Extensive informations can be found at **Wolfram Mathworld** under the keywords:

- Pythagorean Theorem, Heronian Triangle, Orthodiagonal Quadrangle

Among the sites with proofs of the Pythagorean theorem, Alexander Bogomolny's website which contains over 100 proofs, deserves special mention:

- https://www.cut-the-knot.org/pythagoras/

The number of books on the subject is also unmanageable. Here are some recommendations:

- Lietzmann, Walter (1966), *Der pythagoreische Lehrsatz,* 8. Auflage, Teubner, Leipzig (in German)
- Baptist, Peter (1997), *Pythagoras und kein Ende.* Klett, Stuttgart (in German)
- Maor, Eli (2007), *The Pythagorean Theorem: A 4000-Year History,* Princeton, New Jersey: Princeton University Press
- Posamentier, Alfred (2010), *The Pythagorean Theorem: The Story of Its Power and Beauty,* Prometheus Books

Literature on the proofs by dissection:

- Frederickson, Greg (1997), *Dissections: Plane and Fancy,* New York: Cambridge University Press

Among the "miniatures" by Hans Walser there is a wealth of variations on the theme (including proofs by dissection, lunes, Pythagorean butterflies, orthodiagonal quadrilaterals and diagonals, Pythagorean fractals):

- https://www.walser-h-m.ch/hans/Miniaturen_Uebersicht/Pythagoras/index.html

On the subject of "Orthodiagonal squares", please refer to

- Josefsson, Martin: *Characterizations of Orthodiagonal Quadrilaterals,* Forum Geometricorum 12: 13–25 *(2012),* downloadable under https://forumgeom.fau.edu/FG2012volume12/FG201202.pdf

And on the subject of "Heron's triangles that cannot be decomposed" the following recommendation:

- Yiu, Paul (2008), *Heron triangles which cannot be decomposed into two integer right triangles,* 41st Meeting of Florida Section of Mathematical Association of America, downloadable at https://math.fau.edu/yiu/Southern080216.pdf

# General References to Appropriate Literature

The great choice of books on beautiful mathematics makes it difficult to list an appropriate selection.

The collections of mathematical problems originally published in monthly columns of various journals can be stimulating. Here I would like to mention in particular the following authors who have published a great number of books:

- Martin Gardner (among others *Mathematical puzzles and diversions*)
- Theoni Pappas (among others *The Joy of Mathematics*)
- Ian Stewart (among others *Professor Stewart's Cabinet of Mathematical Curiosities*)
- Heinrich Hemme (among others *Heureka!*, in German)

Beautiful and exciting mathematics is the focus of the books by:

- Hans Walser (among others *Der Goldene Schnitt, Geometrische Miniaturen, Symmetrie in Raum und Zeit, DIN A4 in Raum und Zeit* – in German)
- Roger B. Nelsen (among others *Proofs without Words I, II, III*)
- Claudi Alsina and Roger B. Nelsen (among others *Charming proofs, Icons of mathematics*)
- Albrecht Beutelspacher (among others *Wie man in eine Seifenblase schlüpft* – in German)
- Julian Havil (among others *Gamma, Nonplussed!, Impossible?*)
- George G. Szpiro (among others *A mathematical medley: Fifty easy pieces on mathematics*)
- Eli Maor and Eugen Jost (*Beautiful Geometry*)
- Alfred S. Posamentier (among others *Mathematical Amazements and Surprises*)
- Martin Erickson (among others *Beautiful mathematics*)

© Springer-Verlag GmbH Germany, part of Springer Nature 2021  
H. K. Strick, *Mathematics is Beautiful*, https://doi.org/10.1007/978-3-662-62689-4

Among the numerous websites dealing with beautiful mathematics and mathematical puzzles, only the most important ones can be mentioned here:

- www.cut-the-knot.org
- www.gogeometry.com
- www.mathsisfun.com
- www.mathpuzzle.com
- www.recreomath.qc.ca
- www.walser-h-m.ch/hans (mainly in German)
- www.mathematische-basteleien.de (mainly in German)

A collection of very interesting articles on various mathematical topics can be found on the website of *Mathematical Association of America* (MAA). They are especially worth reading because of their historical reference and the question of practicability in school lessons:

- www.maa.org/press/periodicals/convergence

If you want to know more about those personalities who over the centuries have contributed to the development of mathematical theories or have deepened the knowledge about them, I recommend in the first place the website of *MacTutor History of Mathematics archive* the *School of Mathematics & Statistics* at the *University of St Andrews, Scotland:*

- https://mathshistory.st-andrews.ac.uk/Biographies
- For those who would like to get a summary of the life and work of individual selected personalities, we refer to *Heinz Klaus Strick's histories* (https://mathshistory.st-andrews.ac.uk/Strick/) – original version (in German): www.spektrum.de/mathematik/monatskalender/index/

# Index

Printed in the United States
by Baker & Taylor Publisher Services